TEACHER'S MANUAL

PRINCIPLES OF ENGINEERING ECONOMY

EIGHTH EDITION

Eugene L. Grant, Stanford University (Emeritus)
The Late W. Grant Ireson, Stanford University (Emeritus)
Richard S. Leavenworth, University of Florida (Emeritus)

JOHN WILEY & SONS

New York * Chichester * Brisbane * Toronto * Singapore

Copyright © 1990 by John Wiley & Sons, Inc.

This material may be reproduced for testing or instructional purposes by people using the text.

ISBN 0 471 51813 1

Printed in the United States of America

10 9 8 7 6 5 4 3 2 1

PEE Solutions Manual

PREFACE

General Comments

   This instructor's supplement contains the complete answers, and related calculations, for the problems in PRINCIPLES OF ENGINEERING ECONOMY, Eighth Edition. The amount of detail included in each solution varies from problem to problem but, in general, it is the new work in any chapter that is given the most thorough treatment. Although most problems can be formulated and solved by two or more methods, only one is shown here for each problem. It is important for the instructor to point out to students that different methods frequently offer advantages from a computational viewpoint, but that all methods render the same basic results when applied correctly.

   Comments and suggestions have been included in most chapters, and in some cases in individual problems, to assist instructors in making the most effective use of the problem material as a teaching aid.

   Years of teaching engineering economy to a wide variety of students, from college freshmen to practicing engineers with many years of experience, have convinced the authors of the effectiveness of learning through problem solving. Many students who think that they understand the principles of engineering economy proceed to make serious errors when they attempt to apply these principles in solving large, unstructured problems.

   Most of the problems in the text are intentionally simple and well structured, minimizing, to some extent, the time required by the student to solve them. The main objective is to emphasize some particular situation encountered in actual practice. Each problem also affords the instructor the opportunity to illustrate variations of these common situations and to demonstrate the different methods used to solve them.

   In a great many of the problems presented, the numerical answer ultimately is translated into a yes-no, make-buy, repair-replace type of decision. The quantitative analyses that led to these decisions, often prepared by an analyst or engineer, are very important. It is up to the decision maker, however, to apply the necessary judgment to balance the monetary results of alternative courses of action against any non-quantifiable advantages and disadvantages. As part of the overall training of engineers who are or will become decision makers, these problems can serve as the basis for broad discussions on the analysis of alternatives.

   Many of the solutions in this supplement were prepared using electronic calculators and microcomputer programs. Some of these programs are discussed in the next section. To the extent that programs and spreadsheets were used, those solutions may be expected to be more

PEE Solutions Manual   Preface

accurate than those obtained by interpolation from the tables using hand-held or desk-top calculators. This may be particularly true for those problems for which the interest rate is the unknown variable. In using this supplement to check student solutions, emphasis should be put on problem formulation and analysis rather than on the precision of the final numerical answer.

The authors would like to express their sincere appreciation to Mrs. JoAnne Christie Leavenworth for her care and patience in preparing the manuscript of this supplement.

Comments on Computer Programs

On page 72 of the text we discuss the three principal types of computer programs used to solve engineering economy problems. They are: (1) the formula analyzer; (2) the dedicated program; and (3), the generalized spreadsheet. All three of these kinds of programs were used in preparing solutions for this manual.

A computer disk will be provided to each adopter of the text that contains programs and spreadsheet templates for each type of program for use with an IBM-PC or any compatible. The formula analyzer was developed in Turbo Pascal and is listed as EEEE.COM on the disk. The dedicated program is ECON and is written in BASIC. Instructions for running these programs are provided with the disk. **The spreadsheets are for LOTUS 1-2-3.**

The formula analyzer contained on the disk is "An Engineering Economy Expression Evaluator" (**EEEE** Version 5) written by Mr. Thomas Kisko, PE, and is a particularly useful tool for teaching engineering economy. It may be loaded from disk by entering "EEEE". Entering the word "help" or depressing the F1 function key takes the user directly into an excellent Help menu. The program's principal advantage lies in the fact that data and symbols are input to the computer in almost the identical form in which they are formulated on paper. Thus, where we might write

EUAC = $24,000(A/P,8%,10) - $4,500(A/F,8%,10) + $2,000 + $500(A/G,8%,10)

the corresponding computer entry would be

24000*(A/P,8,10) - 4500*(A/F,8,10) + 2000 + 500*(A/G,8,10)

The program may be used to solve directly for i in rate of return calculations. If a problem involves the possibility of multiple roots, a special function call may be used to find all of the real, positive solving roots (values of i). The equation also may be plotted. These functions may be used to find the "zeros" of a wide variety of equations, not just those involving engineering economy symbols. As one word of caution, the continuous compounding factors in the program are not compatible with those given in Tables D-30 and D-31 of the text or with the periodic compounding factors in the program. The factors in EEEE apply to continuous compounding of discrete cash flows at a nominal (APR) interest rate. Those in Tables D-30 and D-31 are the present worth factors for uniform flows, during single periods (D-30) and multiple periods (D-31), with continuous compounding at effective annual interest rates. EEEE is a public domain program. Additional copies and updates may be obtained from the author, Mr. Thomas Kisko, PE, Department of Industrial and Systems Engineering, 303 Weil Hall, University of Florida, Gainesville, FL 32611, for the nominal fee of $10.00.

The dedicated engineering economy program is **ECON** written in BASIC (2.0) by coauthor Richard S. Leavenworth as a teaching tool to accompany the text "Principles of Engineering Economy". It has received extensive classroom testing as well as some application in industry. It is totally interactive and allows the user, following instructions from a sequence of menus, to generate, store, analyze, and perform sensitivity tests on a cash flow series. If an analysis after income taxes is to be made, depreciation charges, taxes, and after income tax cash flows are calculated by a separate program, **ECONTAX**. This program retrieves a table of before tax cash flows, interactively performs the tax calculations based on rules in effect in 1988, and stores the after tax cash flow back on disk where it can be retrieved by ECON and analyzed. There is no doubt but that ECONTAX will have to be revised on a regular basis, a task that is much easier if an entire analysis program does not have to be revised as well. ECON may be used without charge for any educational purpose in support of the text, "Principles of Engineering Economy". Business and industrial usage requires a license from the program author, Richard S. Leavenworth, PO Box 378, Melrose, FL, 32666. An engineering economy tutorial program, "ECOTUTOR", also is available from the program author for those using the text for independent study.

Finally, several examples of what might be called spreadsheet templates for use with LOTUS 1-2-3 are included. We hesitate to call these true templates because they contain none of the formulas required to solve engineering economy problems. Rather, they simply give column headings and initial data entry points for a variety of example applications and problems given in the text. The thought is that spreadsheet problems would be easier to grade if each student submitted work in the same format. However, if formulas were given, very little learning would take place.

We make no claims as to the superiority of these programs over the many others that are available from one source or another. Many instructors have programs of their own, or written by their own students, that they may prefer over those we provide. Some may prefer to limit the amount of student access to computer programs because of the difficulty in policing cheating. Thus the materials we provide are provided to instructors, not directly to students. We will appreciate any comments you have regarding computer programs as well as other aspects of this Instructor's Manual.

Eugene L. Grant
W. Grant Ireson
Richard S. Leavenworth

June, 1989

CONTENTS

Chapter 1:	Basic Principles of Economic Choice	1
Chapter 2:	Equivalence	6
Chapter 3:	Financial Mathematics	11
Chapter 4:	Equivalent Uniform Annual Cash Flow	23
Chapter 5:	Present Worth	31
Chapter 6:	Internal Rate of Return	40
Chapter 7:	Measures Involving Costs, Benefits and Effectiveness	51
Chapter 8:	Some Relationships Between Accounting and Engineering Economy	61
Chapter 9:	Estimating Income Tax Consequences of Certain Decisions	81
Chapter 10:	Increment Costs, Economic Sizing, Sunk Costs, and Interdependent Decisions	99
Chapter 11:	Economy Studies for Retirement and Replacement	108
Chapter 12:	Financing Effects on Economy Studies	127
Chapter 13:	Capital Budgeting and the Choice of a Minimum Attractive Rate of Return	142
Chapter 14:	Prospective Inflation and Sensitivity Analysis	156
Chapter 15:	Use of the Mathematics of Probability in Economy Studies	174
Chapter 16:	Aspects of Economy Studies for Government Activities	180
Chapter 17:	Aspects of Economy Studies for Regulated Businesses	190
Appendix A:	Continuous Compounding of Interest and the Uniform-Flow Convention	204
Appendix B:	Cash flow Series with Two or More Reversals of Sign	209
Appendix C:	The Reinvestment Fallacy in Project Evaluation	217
Appendix F:	Depreciation under the Tax Reform Act of 1986	222

PEE Solutions Manual

CHAPTER 1

Basic Principles of Economic Choice

General Notes:

Chapter 1 delineates ten fundamental concepts of engineering economy that must be thoroughly understood by the practicing engineering economist. These concepts are briefly explained in this chapter, but their significance becomes more clear as one studies the remaining chapters and the Appendixes. The new student to engineering economy needs to start thinking about problems in terms of these concepts; the problems in Chapter 1 are designed to press him/her into this mode of thinking.

It is suggested that instructors emphasize the logic of their methods in discussion of the problems. Students can begin to formulate problems for numerical solution even though the numerical methods are first introduced in the next chapter (Chapter 2.) An incorrectly formulated problem cannot yield a correct solution even though the arithmetic may be perfect.

In non-numerical problems (such as those in Chapter 1), one can expect wide variety in answers to the questions. The value and the validity of a proposed answer can be judged relative to the ten principles listed and the answer's completeness. The following suggestions cover key points in the solution of each problem, but they should not be considered the only points that might logically be considered.

1-1
Some prospective disbursements in connection with car ownership are its initial purchase price, motor fuel and lubricants, tire repair and replacement (if necessary), repairs and maintenance, insurance, license fees and, possibly, parking tickets, traffic fines, and costs associated with accidents. Bus fare is the obvious disbursement in connection with the use of a public bus. Two important irreducibles are the time saved by car ownership (because of no waiting for buses and, presumably, faster transportation to and from campus), and the disadvantage of possibly having no public transportation late in the evening. The major receipt is from the sale of the car at the end of the school year. Conceivably there might also be receipts from passengers who would agree to share expenses. The potential effect of this last benefit on insurance costs should be checked.

PEE Solutions Manual   Chapter 1

1-2
   The disbursement to acquire the car has already been made; it cannot be changed by the choice between keeping the car for the school year and selling it at once. If the car can be sold for the amount that was paid for it, the analysis, in effect, is the same as before. If an immediate sale will bring less than the amount paid for the car, the analysis is the same as in Problem 1-1 except that the net resale price of the car takes the place of its purchase price.

1-3
   (a) $33, the amount obtainable if the ticket should be redeemed by the bus company.
   (b) $50, the amount you will have to pay to replace the ticket.

1-4
   The ranch owner has two major alternatives; build a pole line to the nearest utility system power line or install and operate a generating plant. There are several alternatives under each of these two main alternatives. The kind and sizes of the poles, the size of wire and the spacing of the poles will affect the initialed cost, line losses, and maintenance expenditures. An on-site power generation plant might be fueled by one of several different fuels which in turn will influence the type of turbine, engine, or energy converter needed. Considerable differences will exist among the various alternative courses of action, and will be reflected in differences in the required initial investments, operating costs, and prospective lives of the installations.

   The concepts that need to be discussed are:
   1. Identification and definition of all alternatives.
   2. Estimation of the consquences of each possible alternative.
   3. Can all the differences in the consequences be reduced to monetary terms over time?
   4. Separable decisions, in so far as possible, should be made separately. What is the most economical system? Can it be financed?
   5. Secondary criteria should be examined. How will variations in future needs, and possible variations in fuel costs, affect the decision?
   6. How sensitive is the decision to all of these estimates?

1-5
   Principal alternative courses of action for the boat owner to consider are:
   A. Do nothing now. Continue to operate as in the past, having the engine repaired as necessary.
   The effect of this course of action will be an increasing reduction of receipts due to the increasing rate of breakdowns and loss of goodwill which will accelerate the reduction in income. From recent past experience the owner could estimate the number of days the boat will be "laid up" in the next year due to engine breakdowns. From that, estimates of prospective loss of income and prospective increase in repair expenditures could be made.

B. Have engine completely overhauled now.

This requires an immediate outlay (or investment) of cash and the loss of some number of days of operation while the overhaul takes place. However, the owner could expect to lose fewer days of operation from engine breakdown than would have occurred without the overhaul. Goodwill should improve and the usual engine repair expenditures should be lower than for alternative A. The owner can convert the number of days of operation, having had the overhaul, into an estimate of the additional income to be expected.

C. Purchase a rebuilt engine.
D. Purchase a new engine.

These actions would require larger initial investments than alternative B but should reduce the probability of lost time in the future and probable repair expenses. Both income and goodwill should be improved with a more reliable engine, and the owner should feel more comfortable about passenger safety.

E. Buy a new boat and a new engine.

This alternative will require a much greater investment (cost of new boat and engine less the salvage value of the old boat and engine) but it should reduce the dangers of breakdown and high repair and maintenance expenditures. This should enable operation at nearly 100% of capacity; a new boat also might carry larger parties.

Irreducibles: Goodwill has been mentioned in the discussion of the several alternatives. Goodwill, in this case, means more "repeat calls" for the boat from the same persons; availability is important for repeat business. Good service, availability, and a feeling of safety on the boat are important to prospective customers.

A new engine (or a new boat and engine) will inspire confidence and is likely to result in customers selecting the owner's boat in preference to competing boats. It is very difficult to estimate the increased income that might be generated, so this advantage may best be treated as an irreducible.

1-6

This is valuable as a class discussion problem in which the primary purpose is to identify "whose viewpoint" should be used in its solution. It is obvious that there are several different viewpoints that might be taken including that of the state, the city, residents in the vicinity of the intersection, the general public, and auto insurance companies. What principle should govern the decision? Is the city's viewpoint significantly different from that of the residents of the two streets? Who will pay, in the long run, for the consequences of the different alternatives? Why? How?

1-7

The alternative methods by which a college might provide janitorial service for its buildings are:
  A. Provide its own service by employing:
     1. full-time unskilled labor from the local market
     2. part-time students and full-time supervisors
     3. a combination of these two labor groups

  B. Contract the work to a local janitorial contractor on a fixed price basis.

Some comments regarding the concepts and their application to this problem:

All alternatives need to be identified and carefully defined. The two primary and three subsidiary alternatives listed do not define all the possible subalternatives that might be identified. Alternatives may depend upon some degree of use of full-time and part-time personnel. Each variation will have some effects on expenditures.

The expected consequences of each of the alternatives will include differences in cost as well as in the time at which the functions will be performed. It is assumed that "required quality of service" has been defined and that the consequences of selecting any of the alternatives will not include a difference in the quality of the service.

The primary criterion will be the cost to the college to keep its buildings clean. A secondary criterion involves the uncertainty of the number of persons to be hired under each of the variations of Alternative A. The fact that alternative A-2 or A-3 may provide financial support for needy students might be an irreducible, unless the lack of that work would necessitate the allocation of additional scholarships and assistantships.

The provision of janitorial service is a systems problem. There are many different kinds of activity and work that must be provided, and many different ways of providing each. Looking at the whole "system" and reduction of the work load to quantities of time by labor grade or classification will offer some opportunity for consideration of definitive combinations of people and machines (or mechanical or electrical aids.)

Whose viewpoint presumably is quite clear: the business manager for the college needs to minimize the total cost of providing a service to the college.

1-8

The determination of acceptable alternative ways by which a married student can provide family housing while in college requires first that "acceptable housing" be defined. If the student is male, has children of school age, and a wife who will stay at home most of the time to care for pre-school children, his situation will be much different from a single student, either male or female, seeking to minimize the long run cost. He must set some standards regarding proximity to public schools as well as to the college relative to public transportation or his own car, number of bedrooms, and ability to pay.

The family man or woman has two major alternatives: rent or own. Each of these will also give several alternatives, but they can be summarized by three types of housing: single family units, apartments, mobile homes. Thus, he or she can either buy or rent a single family house, but will probably have to choose from among many available houses. A similar situation may exist for apartments and mobile homes. In the case of rental, he or she may have another class of alternatives represented by rental from the college or rental from private owners.

A decision among alternatives may be influenced by the effect of location on his or her ability to earn supplemental income while going to school, or the availability of jobs for a spouse.

Whose viewpoint obviously is that of the student's entire family. He or she can not make the decision without consideration of the effects of that decision on all members of the family.

The primary criterion is probably to minimize the total cost of housing for the duration of the educational program, with the understanding that the decision must be one that will contribute to the long run happiness of the family. The resources at his or her disposal (for the down payment on a house, for example) and available funds to meet monthly expenses will probably place an upper limit on the set of alternatives from which a choice must be made.

This person should tabulate, very carefully, all of the expenditures and incomes that might arise with each alternative, and also all of the differences that cannot be reduced to monetary terms, such as the fact that one alternative might provide more time to study because it does not involve maintenance (mowing the lawn, shoveling snow, etc.) These irreducibles must be considered in the final decision process and their advantages or disadvantages must be weighed against the monetary differences that exist among the feasible alternatives.

PEE Solutions Manual

CHAPTER 2

Equivalence

2-1 Plan II

End of Year	Interest Due(8%)	Total Before Year-End Payment	Year-End Payment	Owed After Year-End Payment
0				$10,000
1	$800	$10,800	$1,800	9,000
2	720	9,720	1,720	8,000
3	640	8,640	1,640	7,000
4	560	7,560	1,560	6,000
5	480	6,480	1,480	5,000
6	400	5,400	1,400	4,000
7	320	4,320	1,320	3,000
8	240	3,240	1,240	2,000
9	160	2,160	1,160	1,000
10	80	1,080	1,080	0

Plan III

0				$10,000.00
1	$800.00	$10,800.00	$1,490.30	9,309.70
2	744.78	10,054.48	1,490.30	8,564.18
3	685.13	9,249.31	1,490.30	7,759.01
4	620.72	8,379.73	1,490.30	6,889.43
5	551.15	7,440.58	1,490.30	5,950.28
6	476.02	6,426.30	1,490.30	4,936.00
7	394.88	5,330.88	1,490.30	3,840.58
8	307.25	4,147.83	1,490.30	2,657.50
9	212.60	2,870.10	1,490.30	1,379.80
10	110.38	1,490.18	1,490.18*	0

*Difference due to rounding error in (A/P) factor.

Total repayment	8% interest	9% interest
Plan II	$14,000	$14,950
Plan III	14,902.88	15,580

PEE Solutions Manual   Chapter 2

2-2 Plan I

End of Year	Interest Due (15%)	Total Before Year-End Payment	Year-End Payment	Owed After Year-End Payment
0				$12,000
1	$1,800	$13,800	$1,800	12,000
2	1,800	13,800	1,800	12,000
3	1,800	13,800	1,800	12,000
4	1,800	13,800	13,800	0

Plan II

0				$12,000
1	$1,800	$13,800	$4,800	9,000
2	1,350	10,350	4,350	6,000
3	900	6,900	3,900	3,000
4	450	3,450	3,450	0

Plan III

0				$12,000.00
1	$1,800.00	$13,800.00	$4,203.24	9,596.76
2	1,439.51	11,036.27	4,203.24	6,833.03
3	1,024.96	7,857.99	4,203.24	3,654.75
4	548.21	4,202.96	4,202.96*	0

*Difference due to rounding error in (A/P) factor.

Plan IV

0				$12,000.00
1	$1,800.00	$13,800.00	0.00	13,800.00
2	2,070.00	15,870.00	0.00	15,870.00
3	2,380.50	18,250.50	0.00	18,250.50
4	2,737.58	20,988.08	$20,988.08	0

2-3 Repayment of $6,000 in 4 years at 12%.

End of Year	Interest Due (12%)	Total Before Year-End Payment	Year-End Payment	Owed After Year-End Payment
0				$6,000
1	$720	$6,720	$2,220	4,500
2	540	5,040	2,040	3,000
3	360	3,360	1,860	1,500
4	180	1,680	1,680	0

The total amount of principal and interest repaid is $7,800.

PEE Solutions Manual   Chapter 2

2-3 (cont.) Repayment of $6,000 in 10 years at 12%.

End of Year	Interest Due (12%)	Total Before Year-End Payment	Year-End Payment	Owed After Year-End Payment
0				$6,000
1	$720	$6,720	$1,320	5,400
2	648	6,048	1,248	4,800
3	576	5,376	1,176	4,200
4	504	4,704	1,104	3,600
5	432	4,032	1,032	3,000
6	360	3,360	960	2,400
7	288	2,688	888	1,800
8	216	2,016	816	1,200
9	144	1,344	744	600
10	72	672	672	0

The total amount of principal and interest repaid is $9,960.

2-4 Repayment of $5,000 in 5 years at 8%.

0				$5,000
1	$400	$5,400	$1,400	4,000
2	320	4,320	1,320	3,000
3	240	3,240	1,240	2,000
4	160	2,160	1,160	1,000
5	80	1,080	1,080	0

The total amount of principal and interest repaid is $6,200.

Repayment of $5,000 in 5 years at 15%.

0				$5,000
1	$750	$5,750	$1,750	4,000
2	600	4,600	1,600	3,000
3	450	3,450	1,450	2,000
4	300	2,300	1,300	1,000
5	150	1,150	1,150	0

The total amount of principal and interest repaid is $7,250.

2-5 Plan I - Repayment of $2,000 in 4 years at 8%.

0				$2,000
1	$160	$2,160	$160	2,000
2	160	2,160	160	2,000
3	160	2,160	160	2,000
4	160	2,160	2,160	0

Plan II

0				$2,000
1	$160	$2,160	$660	1,500
2	120	1,620	620	1,000
3	80	1,080	580	500
4	40	540	540	0

PEE Solutions Manual					Chapter 2

2-5 (cont.)   Plan III

End of Year	Interest Due (8%)	Total Before Year-End Payment	Year-End Payment	Owed After Year-End Payment
0				$2,000.00
1	$160.00	$2,160.00	$603.84	1,556.16
2	124.49	1,680.65	603.84	1,076.81
3	86.14	1,162.95	603.84	559.11
4	44.73	603.84	603.84	0

Plan IV

0				$2,000.00
1	$160.00	$2,160.00	$0.00	2,160.00
2	172.80	2,332.80	0.00	2,332.80
3	186.62	2,519.42	0.00	2,519.42
4	201.55	2,720.97	2,720.97	0

	Plan I	Plan II	Plan III	Plan IV
Comparative Total Repayments	$2,640	$2,400	$2,415.36	$2,720.97

2-6 Repayment of $8,000 in 5 years at 15%; Plan II.

0				$8,000
1	$1,200	$9,200	$2,800	6,400
2	960	7,360	2,560	4,800
3	720	5,520	2,320	3,200
4	480	3,680	2,080	1,600
5	240	1,840	1,840	0

2-7 Repayment of $8,000 in 5 years at 15%, Plan III.

0				$8,000.00
1	$1,200.00	$9,200.00	$2,386.56	6,813.44
2	1,022.02	7,835.46	2,386.56	5,448.90
3	817.34	6,266.24	2,386.56	3,879.68
4	581.95	4,461.63	2,386.56	2,075.07
5	311.26	2,386.33	2,386.33*	0

*Difference due to rounding error in (A/P) factor.

2-8 Repayment of $10,000 in 5 years at 9%.
Plan II

0				$10,000
1	$900	$10,900	$2,900	8,000
2	720	8,720	2,720	6,000
3	540	6,540	2,540	4,000
4	360	4,360	2,360	2,000
5	180	2,180	2,180	0

PEE Solutions Manual   Chapter 2

2-8 (cont.) Plan III

End of Year	Interest Due (9%)	Total Before Year-End Payment	Year-End Payment	Owed After Year-End Payment
0				$10,000.00
1	$900.00	$10,900.00	$2,570.90	8,329.10
2	749.62	9,078.72	2,570.90	6,507.82
3	585.70	7,093.52	2,570.90	4,522.62
4	407.04	4,929.66	2,570.90	2,358.76
5	212.29	2,571.05	2,571.05*	0

*Difference due to rounding error i (A/P) factor.

Total Payments	From Table 2-1 (10 yrs)	From above
Plan II	$14,950	$12,700
Plan III	$15,580	$12,855

PEE Solutions Manual

CHAPTER 3

Financial Mathematics

3-1

Semiannually: $i_e = (1+0.12/2)^2 - 1 = (1.06)^2 - 1 = 0.1236 =$ <u>12.36%</u>

Quarterly: $i_e = (1+0.12/4)^4 - 1 = (1.03)^4 - 1 = 0.1255 =$ <u>12.55%</u>

Monthly: $i_e = (1+0.12/12)^{12} - 1 = (1.01)^{12} - 1 = 0.1268 =$ <u>12.68%</u>

3-2

Nominal $= 12(1.5\%) =$ <u>18%</u>.

Effective $= (1.015)^{12} - 1 = 0.1956 =$ <u>19.56%</u>

3-3

$i_e = (1+0.15/12)^{12} - 1 = (1.0125)^{12} - 1 = 0.16075 =$ <u>16.075%</u>

By logarithms: 12 log 1.0125 = 12(0.005395)

= 0.06474; $\log^{-1} 0.06474 = 1.16075$

$i_e = 1.16075 - 1 = 0.16075 =$ <u>16.075%</u>

3-4

Daily: $i_e = (1+0.15/365)^{365} - 1 = 1.16180 - 1 = 0.16180$

= <u>16.180%</u>. By logarithms:

365 log (1+0.15/365) = 365(0.0001784) = 0.06513

$i_e = \log^{-1} 0.06513 - 1 = 0.16180 =$ <u>16.180%</u>

Continuous: $i_e = e^r - 1 = e^{0.15} - 1 = 1.16183 - 1 =$ <u>0.16183</u>

= <u>16.183%</u>.

3-5

P = $500(P/A,5%,20) + $100(P/G,5%,20)

= $500(12.462) + $100(98.488) = <u>$16,080</u>

11

3-6

P = $2,400(P/A,5%,20) - $100(P/G,5%,20)

= $2,400(12.462) - $100(98.488) = $20,060

3-7

The semiannual payment is $100,000(A/P,3.5%,60)

= $100,000(0.04009) = $4,009

40 payments remain after 20 have been made, thus the principal remaining is P_{20} = $4,009(P/A,3.5%,40) = $4,009(21.355)

= $85,612. $100,000 - $85,612 = $14,388 of the original principal has been retired.

3-8

(a) (A/P,i%,10) = $1,300/$8,000 = 0.1625
    (A/P,9%,10) = 0.15582; (A/P,10%,10) = 0.16275
    by interpolation, i = 9.96% ≈ 10%

(b) (A/P,i%,20) = $680/$6,000 = 0.11333
    (A/P,9%,20) = 0.10955; (A/P,10%,20) = 0.11746
    by interpolation, i = 9.5%

(c) (A/P,i%,15) = $400/$6,000 = 0.06667 = 1/n
    Therefore i = 0%

(d) A = 0 -$10,000(A/P,i%,20)+$300+$50(A/G,i%,20)
    Try 3.5%; -$10,000(0.07036)+$300+$50(8.365) = +$14.65
    Try 4.0%; -$10,000(0.07358)+$300+$50(8.209) = -$25.35
    by interpolation, i = 3.7%

(e) Since the capital recovery factor (A/P) approaches i as n approaches infinity, A = Pi from which i = $225/$5,000 = 0.045, or i = 4.5%

3-9

The differences between the two plans, purchase vs. lease, are the purchase and resale values vs. the beginning-of-year lease payments. The interest rate that will make the present value (for example) of these two plans equal is the prospective rate of return on the investment in the land purchase. Thus equate:
-$80,000 + $100,000(P/F,i%,15) = -$5,000[1 + (P/A,i%,14)]
at i = 7%; -$75,000 + $100,000(0.3624)+$5,000(8.245) = $4,965
at i = 8%; -$75,000 + $100,000(0.3152)+$5,000(8.244) = -$2,260
by interpolation, i = 7.7%

PEE Solutions Manual   Chapter 3

3-10
   This problem can be visualized as substituting $55,000 - $5,000 now for $5,000 at the end of each of the next 14 years. Thus (A/P,i%,14) = $5,000/$50,000 = 0.10000. By interpolation, <u>i = 4.8%</u>. It may be interpreted as the interest rate that would be paid on a loan of $50,000 if the decision is made not to prepay the rent.

3-11
   (A/P,i%,16) = $14.44/$200 = 0.07220

   (A/P,1.5%,16) = 0.07077; (A/P,2%,16) = 0.07365;

by interpolation, i = <u>1.75%</u> per week. This is a nominal 90.91% per year and an effective $(1+0.01748)^{52} - 1 = 146.3\%$ per year.

3-12
   The purchaser must be able to sell the land 10 years hence for the compound amount of all expenditures up to that time. Thus the selling price must be:
   F = $20,000(F/P,15%,10) + $400(F/A,15%,10)
     + $40(A/G,15%,10)(F/A,15%,10)
     = $20,000(4.0456) + $400(20.304) + $40(3.383)(20.304)
     = $80,912 + $8,122 + $2,747 = <u>$91,781</u>

3-13
   Since some of the payments are intermittent, they must be converted to an equivalent P(or F) before conversion to a uniform A for 30 years all at 9% interest.
   P = $10,000[1 + (P/F,9%,10) + (P/F,9%,20)]
     = $10,000(1 + 0.4224 + 0.1784) = $16,008
   A = $16,008(A/P,9%,30) + 2,000 = $16,008(0.09734)
     + $2,000 = <u>$3,558.22</u>

3-14
   A formula to relate a series of <u>beginning of period</u> deposits, B, in a sinking fund for n periods with interest at i per period to the amount accumulated, F, at the end of the nth period, may be derived directly, or the end-of-period formulas given in Chapter 3 may be used to derive it.

   F can be expressed as the summation of a series:

   $F = B[(1+i)+(1+i)^2 + \text{-----} + (1+i)^{n-1} + (1+i)^n]$

   Multiplying both sides of the equation by (1+i), we have

   $F(1+i) = B[(1+i)^2 + (1+i)^3 + \text{-----} + (1+i)^n + (1+i)^{n+1}]$

13

PEE Solutions Manual   Chapter 3

3-14 (cont)
Subtracting the second equation from the first, we obtain

$$F - F(1+i) = B[(1+i) - (1+i)^{n+1}]$$

That equation can be cleared to

$$F = B\frac{[(1+i) - (1+i)^{n+1}]}{1 - (1+i)} = B\frac{(1+i)[1-(1+i)^n]}{1 - (1+i)} \text{ ; from which:}$$

$$B = F\frac{1 - (1+i)}{[1-(1+i)^n](1+i)} = F\frac{i}{[(1+i)^n - 1](1+i)} \text{ .}$$

3-15
Let B' be the end-of-period deposit in a sinking fund that will accumulate to F, at interest = i, in n periods. The sinking fund factor multiplied by F will give B':

$$B' = \frac{i}{(1+i)^n - 1} F \quad ;$$ At each period the end-of-period amount B' is equal to the beginning-of-period amount B(F/P,i%,n), or to express B in terms of B', B=B'(P/F,i%,1)

Substituting the formula for (P/F,i%,n) in the equation B', we have

$$B = \frac{i}{(1+i)^n - 1} \times \frac{1}{(1+i)} F = (A/F,i\%,n)(P/F,i\%,1) F$$

From the formula in 3-14:

$$B = \frac{0.20}{[(1+0.20)^{10}-1](1+0.20)} F = \underline{0.032102\ F}$$

From Table D-23:   B = 0.03852(0.8333)   F = <u>0.032099 F</u>

3-16
(F/A,12%,6) = 8.115 from Table D-17.

(a)   (F/A,12%,6) = 1 + $\sum_{t=1}^{9}$ (F/P,12%,t)

  = 1 + 1.12+1.2544+1.4049+1.5735+1.7623 = 8.1151 = <u>21.3214</u>

(b)   (F/A,i%,n) = 1/(A/F,i%,n)

  = 1/(A/F,12%,6) = 1/0.12323 = <u>8.1149</u>

14

3-17
(A/P,i%,n); (A/P,10%,6) = 0.22961 from Table D-15.

(a) $(A/P,i\%,n) = [\sum_{t=1}^{n} (P/F,i\%,t)]^{-1}$

(A/P,10%,6) = [0.9091+0.8264+0.7513+0.6830+0.6209
+ 0.5645]$^{-1}$ = [4.3552]$^{-1}$ = <u>0.22961</u>

(b) (A/P,i%,n) = 1/(P/A,i%,n)

(A/P,10%,6) = 1/(4.355) = <u>0.22962</u>

3-18
(a) (A/P,i%,20) = $750/$5,000 = 0.15000
(A/P,13%,20) = 0.14235: (A/P,14%,20) = 0.15099.
By interpolation: i% = <u>13.9%</u>

(b) (A/P,i%,10) = $800/$6,000 = 0.13333
(A/P,5%,10) = 0.12950; (A/P,6%,10) = 0.13267.
By interpolation: i% = <u>5.6%</u>

(c) $50,000(A/P,i%,15) = $9,000 + $12,000(A/F,i%,15)
@13%: -$60,000(0.15474) + $9,000 + $12,000(0.02474) = +$12.48
@14%: -$60,000(0.16281) + $9,000 + $12,000(0.02281) = -$494.88
By interpolation: i% = <u>13.0%</u>

3-19
(a) $4,000(P/F,8%,12) = $4,000(0.3971) = <u>$1,588.40</u>

(b) $12,000(A/P,8%,20) = $12,000(0.10185) = <u>$1,222.20</u>

(c) $150,000((P/F,8%,40) = $150,000(0.0460) = <u>$6,900.00</u>

(d) A = Pi = $250,000(0.08) = <u>$20,000</u>

3-20
A = [$15,000 + $5,000(P/F,12%,5)](A/P,12%,10) + $1,000
+ $500(A/G,12%,10) = [$15,000 + $5,000(0.5674)](0.17698)
+ $500(3.585) = $3,156.79 + $1,000 + $1,792.50 = <u>$5,949.29</u>

3-21

$(1+i) = e^r$. Thus: $r = \ln(1+i)$.
At 15%: $r = \ln(1 + 0.15) = \ln(1.15)$ = <u>0.139762</u> or <u>13.9762%</u>

15

PEE Solutions Manual   Chapter 3

3-22

$(1 + r/365)^{365} = (1 + i_e)$.

$365 \log(1 + r/365) = \log(1.15) = 0.06069784$

$\log^{-1}(1 + r/365) = \log^{-1}(0.06069784/365) = 1.000382983$

$= 1 + r/365$ from which r = **0.139789** or **13.9789%**

3-23

$(A/P, i\%, n) = [i(1+i)^n]/[(1+i)^n - 1]$

n=5: $(A/P, 28\%, 5) = [0.28(1.28)^5]/[1.28^5 - 1]$
       $= 0.9621/2.4360 = \underline{0.39494}$

From Appendix D:
   $(A/P, 25\%, 5) = 0.37185$; $(A/P, 30\%, 5) = 0.41058$.
By linear interpolation for 28%,
   $(A/P, 28\%, 5) = 0.37185 + (3/5)(0.41058 - 0.37185)$
   $= 0.39509$. This differs from the correct value by +0.038%.

n=10: $(A/P, 28\%, 10) = [0.28(1.28)^{10}]/[1.28^{10} - 1]$
        $= 3.30566/10.80592 = \underline{0.30591}$

From Appendix D:
   $(A/P, 25\%, 10) = 0.28007$; $(A/P, 30\%, 10) = 0.32346$.

By linear interpolation for 28%,
   $(A/P, 28\%, 10) = 0.28007 + (3/5)(0.32346 - 0.28007)$
   $= 0.30610$. This differs from the correct value +0.062%.

n=20: $(A/P, 28\%, 20) = [0.28(1.22)^{20}]/[1.28^{20} - 1]$
         $= 39.02630/138.37966 = \underline{0.28202}$

From Appendix D:
   $(A/P, 25\%, 20) = 0.25292$; $(A/P, 30\%, 20) = 0.30159$.

By linear interpolation for 28%,
   $(A/P, 28\%, 20) = 0.25292 + (3/5)(0.30159 - 0.25292)$
   $= 0.28212$. This differs from the correct value by +0.035%.
   In all cases the interpolated value exceeds the actual by a
   very small amount.

PEE Solutions Manual   Chapter 3

3-24

$(P/A, i\%, n) = [(1+i)^n - 1]/[i(1+i)^n]$

<u>n=5</u>: $(P/A, 28\%, 5) = [(1.28)^5 - 1]/[0.28(1.28)^5]$
= 2.4360/0.9621 = <u>2.5320</u>

From Appendix D:
$(P/A, 25\%, 5) = 2.689$; $(P/A, 30\%, 5) = 2.436$.

By linear interpolation for 28%,
$(P/A, 28\%, 5) = 2.689 - (3/5)(2.689-2.436)$
= 2.5372. This differs from the correct value
by +0.205%.

<u>n=10</u>: $(P/A, 28\%, 10) = [(1.28)^{10} - 1]/[0.28(1.28)^{10}]$
= 10.8059/3.3057 = <u>3.2689</u>

From Appendix D:
$(P/A, 25\%, 10) = 3.571$; $(P/A, 30\%, 10) = 3.092$.

By linear interpolation for 28%,
$(P/A, 28\%, 10) = 3.571 - (3/5)(3.571-3.092)$
= 3.2836. This differs from the correct value by
+0.450%.

<u>n=20</u>: $(P/A, 28\%, 20) = [(1.28)^{20} - 1]/[0.28(1.28)^{20}]$
= 138.3797/39.0263 = <u>3.5458</u>

From Appendix D:
$(P/A, 25\%, 20) = 3.954$; $(P/A, 30\%, 20) = 3.316$.

By linear interpolation for 28%,
$(P/A, 28\%, 20) = 3.954 - (3/5)(3.954 - 3.316)$
= 3.5712. This differs from the correct value by
+0.716%. In all cases, the interpolated value exceeds
the actual value. This difference, while increasing,
is less than 1%.

3-25

For n in the range compared (5-20), both factors give higher than actual values when interpolation is used. The (A/P) factor gives closer results (with an error of less than 0.1%) than the (P/A) factor (with an error of less than 1%, an order of magnitude greater).

PEE Solutions Manual   Chapter 3

3-26
n=5: (P/A,10%,5) = 3.791 ; (P/A,25%,5) = 2.689.

By linear interpolation for 15%,
(P/A,15%,5) = 3.791 - (5/15)(3.791-2.689) = **3.4237**

From Table D-20, ((P/A,15%,5) = **3.352**. The interpolated value differs from the actual by +2.14%.

n=10: (P/A,10%,10) = 6.144 ; (P/A,25%,10) = 3.571.

By interpolation for 15%,
(P/A,15%,10) = 6.144 - (1/3)(6.144-3.571) = **5.2863**

From Table D-20, (P/A,15%,10) = **5.019**. The interpolated value differs the actual by +5.33%.

n=20: (P/A,10%,20) = 8.514 ; (P/A,25%,20) = 3.954.

By interpolation for 15%,
(P/A,15%,20) = 8.514 - (1/3)(8.514-3.954) = **6.9940**

From Table D-20, (P/A,15%,20) = **6.259**. The interpolated value differs from the actual by +11.74%. The percentage error increases rapidly as n increases.

3-27
F = $24(F/P,6%,369) = $24(1.06)^{369} = $52,248,700,000. This sum has been said to approximate the value of Manhattan today.

3-28
P = $3,000(P/F,15%,5) + $8,000(P/F,15%,10)

+$15,000[(P/F,15%,15) + (P/F,15%,20)]

= $3,000(0.4972) + $8,000(0.2472)

+ $15,000(0.1229 + 0.0611) = $6,229.20

A = $6,229.2(A/P,15%,25) = $6,229.2(0.15470) = **$963.66**

PEE Solutions Manual   Chapter 3

3-29
In order to solve this problem, you need to find the value of $(1+i_e)$ compounded daily. You should assume that the deposits are made at the beginning of each year for 20 years with the final withdrawl made at the end of the 20 year period. Thus you want:
$F = A(F/A, i_e\%, 20)(1+i_e)$. The value of the (F/A) factor may be found directly or interpolated from the tables.

$(1 + i_e) = (1 + 0.06/365)^{365} = (1.06)^{365} = 1.06183$.

$(F/A, i_e\%, 20) = [1.06183^{20} - 1]/[1.06183 - 1] = 37.5180$

$F = \$1,000(37.5180)(1.06183) = \underline{\$39,837.83}$

Compounded annually, the final amount would have been:
$F = \$1,000(F/A, 6\%, 20)(1 + 0.06) = \$1,000(36.786)(1.06) = \underline{\$38,939.16}$
This is a difference of +2.17% in total earnings.

3-30
Assuming a constant rate per year of price level change, f%, is equivalent to using a single payment compound amount factor at that rate. That is, if R is the base price in the year an estimate is made and that price is expected to grow at the rate f%, then the price for any future year n is:

$R(1 + f)^n = R(F/P, f\%, n)$

and the single payment compound amount factor may be used to find the price by substituting f% for i% in the appropriate table.
Assuming an 8% increase per annum in the price of gasoline, the price per gallon in 5, 10, and 20 years, respectively, will be:
  $\$1.12(F/P, 8\%, 5) = \$1.12(1.4693) = \$1.65$
  $\$1.12(F/P, 8\%, 10) = \$1.12(2.1589) = \$2.42$
  $\$1.12(F/P, 8\%, 20) = \$1.12(4.6610) = \$5.22$
The unfortunate fact in assuming a geometric growth in any number is that eventually the number goes to infinity. No cost factor grows in such a manner. As will be seen in Chapter 14, inflation increases (or decreases) by rates that vary each year. Thus the assumption that price increases will occur at some constant rate per year is bound to lead to errors. For purposes of calculation, the assumption may be useful over the short haul but is not very useful over the long haul.

3-31
$CR = \$75,000(A/P, 16\%, 10) - \$6,000(A/F, 16\%, 10)$
$= \$75,000(0.20690) - \$6,000(0.04690)$
$= \$15,517.50 - \$281.40 = \underline{\$15,236.10}$

PEE Solutions Manual   Chapter 3

3-32
 (A/P,i%,15) = $51/$600 = 0.0850.
 (A/P,3%,15) = 0.08377; (A/P,3.5%,15) = 0.08683
 by interpolation, i = 3.2% per month. The nominal interest rate per year is 12(3.2%) = <u>38.4%</u>. The effective interest rate per year is:

$$(1 + 0.032)^{12} - 1 = \underline{46.0\%}$$

3-33
 We may use the formula: $F = P(1 + i)^n = P(F/P,i\%,n)$ and solve for i.

 $2,689.37 = $1,000(1 + i)^{18}$
 18log(1 + i) = log(2,689.37/1,000) = 0.4296506
 log(1 + i) = 0.023869
 i = antilog 0.023869 - 1 = <u>0.0564999 or 5.65%</u>
 Interpolation in Appendix D between 5.5% and 6.0% leads to 5.646%.

3-34
 There are 20(4) = 80 compounding periods in this problem at 6%/4 = 1.5% per period. Thus:
  F = $5,000(F/P,1.5%,80) = $5,000(3.2907) = <u>$16,453.50</u>
 If interest were compounded annually, F = $5,000(F/P,6%,20) = $5,000(3.2071) = <u>16,035.50</u> which is 2.54% less earnings than by quarterly compounding.

3-35
 First find the amount in the account at the end of 10 years.
  F(10) = $5,000(F/P,1.5%,40) = $5,000(1.8140) = $9,070.
 From this, F(20) = ($9,070 - $4,000)(F/P,1.5%,20) = $5,070(1.8140)
  = <u>$9,196.98</u>

3-36
 20 years: $50,000(A/P,11%,20) = $50,000(0.12558) = <u>$6,279</u>

 30 years: $50,000(A/P,11%,30) = $50,000(0.11502) = <u>$5,751</u>

 The difference is only $528 per year.

3-37
 20 - yr mortgage: $6,279(P/A,11%,5) = $6,279(3.696) = <u>$23,207</u>

 30 - yr mortgage: $5,751(P/A,11%,15) = $5,751(7.191) = <u>$41,355</u>

20

PEE Solutions Manual   Chapter 3

3-38
$40(1+i) = 50; $i = (\$50/\$40) - 1 = \underline{25\% \text{ per week}}$.

$52(25\%) = 1,300\%$ per year nominal;

$(1.25)^{52} - 1 = 10,947,544\%$ per year effective.

3-39
$F = \$3,000(F/P,1.5\%,12) + \$200(F/A,1.5\%,12)$

$\quad + \$20(A/G,1.5\%,12)(F/A,1.5\%,12)$

$\quad = \$3,000(1.1956) + \$200(13.041) + \$20(5.323)(13.041) = \underline{\$7,583.34}$

$A_{12} = 11(\$20) + \$200 = \underline{\$420}$

3-40
$\$750 = \$50(P/A,i\%,8) + \$20(P/G,i\%,8) + \$1,500(P/F,i\%,8)$

@18%; $-\$750 + \$50(4.078) + \$20(10.829) + \$1,500(0.2660) = +\$69.48$

@20%; $-\$750 + \$50(3.837) + \$20(9.883) + \$1,500(0.02326) = -\$11.59$

by interpolation, $i = \underline{19.7\%}$

3-41
$(A/P,i\%,24) = \$125/2,250 = 0.05556$

$(A/P,2\%,24) = 0.05287$; $(A/P,2.5\%,24) = 0.05591$;

by interpolation, $\underline{i = 2.44\%}$ per month. This is $12(2.44)$
$= 29.25\%$ nominal per year and $(1.0244)^{12} - 1 = 33.55\%$
effective per year.

3-42
$(A/P,i\%,24) = \$304.07/\$6,000 = 0.05068$

$(A/P,1.5\%,24) = 0.04992$; $(A/P,2\%,24) = 0.05287$;

by interpolation, $\underline{i = 1.629\%}$ per month, which is nominally

$12(1.629) = 19.54\%$ and effectively $(1.01629)^{12} -1 = 21.39\%$
per year. The insurance should be considered part of the
financing because it is required by the lender to protect the lender.

21

PEE Solutions Manual   Chapter 3

3-43
P_{65} = $6,000(P/A,2%,100) = $6,000(43.098) = $258,588

(a) A_{25} = $258,588(A/F,2%,160) = $258,588(0.000878) = $227.13

(b) A_{40} = $258,588(A/F,2%,100) = $258,588(0.00320) = $827.48

3-44
P_{65} = $6,000/0.02 = $300,000

(a) A_{25} = $300,000(A/F,2%,160) = $300,000(0.000878) = $263.40

(b) A_{40} = $300,000(A/F,2%,100) = $300,000(0.00320) = $960.00

The deposits would increase by 16% in each case.

3-45
Since the deposits after the first are to be made every 4th interest period, they can be converted to equivalent uniform period amounts by $1,000(A/F,1.5%,4) = $1,000(0.24444) = $244.44

Then, F = $1,000(F/P,1.5%,80) + $244.44(F/A,1.5%,80)

= $1,000(3.2907) + $244.44(152.711) = $40,619

3-46
Down payment, company financing, is $2,500.
Monthly payment is $10,000(A/P,0.325%,48)
  = $10,000(0.022534) = $225.34.
With credit union financing, down payment is $3,500.
Monthly payment is $9,000(A/P,0.6269%,48)
  = $9,000(0.024189) = $217.71.
By foregoing the dealer's promotional financing, you save $1,000 on the down payment thus reducing your monthly payments by $7.63 even though the interest rate is substantially higher.

PEE Solutions Manual

CHAPTER 4

Equivalent Uniform Annual Cash Flow

4-1

	Plan I	Plan II
Interest on land investment		
$75,000(0.09) =$	$6,750	
$100,000(0.09) =$		$9,000
CR on structures		
$75,000(A/P,9\%,30) = \$75,000(0.09734) =$	7,301	
$150,000(A/P,9\%,50)-\$30,000(A/F,9\%,50)$		
$= \$150,000(0.09123)-\$30,000(0.00123) =$		13,648
Equivalent uniform annual disbursements		
$7,000+\$3,000(P/A,9\%,10)0.09$		
$= \$7,000+\$3,000(6.418)0.09 =$	8,733	
$4,000		4,000
	$22,784	$26,648

4-2

	Type A	Type B
CR for Pipe		
$120,000(A/P,6\%,60) = \$120,000(0.06188) =$	$7,426	
$80,000(A/P,6\%,30) = \$80,000(0.07265) =$		$5,812
CR for Pumps		
$15,000(A/P,6\%,20) = \$15,000(0.08718) =$	1,308	
$20,000(A/P,6\%,20) = \$20,000(0.08718) =$		1,744
Equivalent Uniform Annual Energy Cost		
$3,000+\$60(A/G,6\%,60)$		
$= \$3,000+\$60(14.791) =$	3,887	
$4,000+\$80(14.791) =$		5,183
	$12,621	$12,739

4-3

		Concrete
CR = $350,000(A/P,7\%,90) = \$350,000(0.07016) =$		$24,556
Annual Upkeep		2,500
		$27,056

		Wood
CR:		
Earth fill:	$50,000(A/P,7\%,90) = \$50,000(0.07016) =$	$3,508
Bleachers:	$100,000(A/P,7\%,30) = \$100,000(0.08059) =$	8,059
Seats:	$40,000(A/P,7\%,15) = \$40,000(0.10979) =$	4,392
Painting:	$10,000(A/P,7\%,3) = \$10,000(0.38105) =$	3,810
		$19,769

PEE Solutions Manual   Chapter 4

4-4

		Type Y	Type Z
CR:			
$8,400(A/P,16%,6) = $8,400(0.27139) =		$2,280	
$10,800(A/P,16%,9) = $10,800(0.21708) =			$2,344
Annual operating costs		$1,700	1,500
Extra annual income taxes			200
		$3,980	$4,044

4-5

		4-Years	5-Years
CR:			
$8,400(A/P,10%,4)-$2,400(A/F,10%,4)			
=$8,400(0.31547)-$2,400(0.21547) =		$2,133	
$8,400(A/P,10%,5)-$1,500(A/F,10%,5)			
=$8,400(0.26380)-$1,500(0.16380) =			$1,970
Operation & Maintenance:			
$2,400+$400(A/G,10%,4)=$2,400+$400(1.381) =		2,952	
$2,400+$400(A/G,10%,5)=$2,400+$400(1.810) =			3,124
		$5,085	$5,094

4-6

		Mach. J	Mach. K
CR:			
$50,000(A/P,12%,12)-$14,000(A/F,12%,12)			
=$50,000(0.16144)-$14,000(0.04144) =		$7,492	
$30,000(A/P,12%,12) = $30,000(0.16144) =			$4,843
Operation & Maintenance:			
$6,000+$300(A/G,12%,12) = $6,000+$300(4.190) =		7,257	
$600+$80(4.190) =		935	
$8,000+$500(4.190) =			10,095
		$15,684	$14,938

4-7

		Concrete	Frame
CR:			
$1,160,000(A/P,9%,50) = $1,160,000(0.09123) =		$105,827	
$480,000(A/P,9%,25) = $480,000(0.10181) =			$48,869
Operation & Maintenance:		4,000	7,500
Insurance:$1.50[$4,000,000+(0.75)$1,160,000]/$1,000 =		7,305	
$4[$4,000,000 + (0.75)$480,000]/$1,000 =			17,440
Property Taxes: 0.015($1,160,000) =		17,400	
0.015($480,000) =			7,200
Extra Annual Income Tax		11,642	
		$146,174	$81,009

24

PEE Solutions Manual   Chapter 4

4-8
  CR: Concrete   Frame
  $1,160,000(A/P,18%,50) = $1,160,000(0.18005) = $208,858
  $480,000(A/P,18%,25) = $480,000(0.18292) = $87,802
  Operation & Maintenance   4,000   7,500
  Insurance (See problem 4-7.)   7,305   17,440
  Property Taxes   17,400   7,200
                                                        $237,563 $119,942

4-9
  CR: Model H   Model K
  $12,000(A/P,20%,3)-$2,500(A/F,20%,3)
  = $12,000(0.47473)-$2,500(0.27473) = $5,010
  $16,000(A/P,20%,4)-$2,500(A/F,20%,4)
  = $16,000(0.38629)-$2,500(0.18629) = $5,715
  Operation & Maintenance
  $5,000+$300(A/G,20%,3) = $5,000+$300(0.879) =   5,264
  $3,800+$200(A/G,20%,4) = $3,800+$200(1.274) =   4,055
                                                         $10,274 $9,770

4-10
  CR: Unit A   Unit B
  $40,000(A/P,14%,8)=$40,000(0.21557) = $8,623
  $50,000(A/P,14%,10)-$5,000(A/F,14%,10)
  =$50,000(0.19171)-$5,000(0.05171) = $9,327
  Annual disbursements   1,800   1,200
  Add'l income tax   440
                                                         $10,423 $10,967

4-11
  CR: Unit A   Unit B
  $40,000(A/P,25%,8) = $40,000(0.30040) = $12,016
  $50,000(A/P,25%,10)-$5,000(A/F,25%,10)
  = $50,000(0.28007)-$5,000(0.03007) = $13,853
  Annual disbursements   1,800   1,200
                                                         $13,816 $15,053

4-12
  CR: System E   System F
  $44,000(A/P,15%,6)-$2,000(A/F,15%,6)
  = $44,000(0.26424)-$2,000(0.11424) = $11,398
  $72,000(A/P,15%,10)-$6,000(A/F,15%,10)
  = $72,000(0.19925)-$6,000(0.04924) = $14,050
  Annual Operation & Maintenance   16,800   9,800
  Extra annual income tax   2,960
                                                        $28,198 $26,810

25

PEE Solutions Manual   Chapter 4

4-13
    CR: System E   System F
    $44,000(A/P,30%,6)-$2,000(A/F,30%,6)
    = $44,000(0.37839)-$2,000(0.07839) = $16,492
    $72,000(A/P,30%,10)-$6,000(A/F,30%,10)
    = $72,000(0.32346)-$6,000(0.02346) = $23,148
    Annual Operation & Maintenance   16,800   9,800
                                                       $33,292 $32,948

4-14
    CR: Unit X   Unit Y
    $16,000(A/P,20%,10)-$2,000(A/F,20%,10)
    = $16,000(0.23852)-$2,000(0.03852) = $3,739
    $24,000(0.23852)-$3,000(0.03852) = $5,609
    Annual costs: 9,000   7,000
    $30(A/G,20%,10) = $30(3.074) =   92
    Inc. tax = $520+$12(3.074) =   557
                                                       $12,831 $13,166

4-15 $P(A/P,i\%,n) - S(A/F,i\%,n)$

$$= P[i(1 + i)^n]/[(1 + i)^n - 1] - Si/[(1 + i)^n - 1]$$

(a) $(P - S)(A/P,i\%,n) + Si$

$$= (P - S)[i(1 + i)^n]/[(1 + i)^n - 1] + Si$$

however: $Si\{1 - (1 + i)^n/[(1 + i)^n - 1]\}$

$$= Si[(1 + i)^n - 1 - (1 + i)^n]/[(1 + i)^n - 1]$$

$$= -Si/[(1 + i)^n - 1], \text{ and therefore:}$$

$$= P[i(1 + i)^n]/[(1 + i)^n - 1] - Si/[(1 + i)^n - 1]$$

(b) $[P - S(P/F,i\%,n)](A/P,i\%,n)$

$$= P[i(1 + i)^n]/[(1 + i)^n - 1]$$

$$- S[(1 + i)^{-n}][i(1 + i)^n]/[(1 + i)^n - 1]$$

$$= P[i(1 + i)^n]/[(1 + i)^n - 1] - Si/[(1 + i)^n - 1]$$

PEE Solutions Manual   Chapter 4

(c) $P_i + (P - S)(A/F, i\%, n)$

   $= P_i + P_i/[(1 + i)^n - 1] - S_i/[(1 + i)^n - 1]$

   however: $P_i\{1 + 1/[(1 + i)^n - 1]\}$

   $= P_i[(1 + i)^n - 1 + 1]/[(1 + i)^n - 1]$, and therefore:

   $= P[i(1 + i)^n]/[(1 + i)^n - 1] - S_i/[(1 + i)^n - 1]$

4-16
```
CR:                                              Type A      Type B
$8,500(A/P,13%,12)-$850(A/F,13%,12)
= $8,500(0.16899)-$850(0.03899) =                $1,403
$11,000(0.16899)-$850(0.03899) =                             $1,826
Annual Operation & Maintenance                    2,550       2,125
Extra Annual Income Tax                                          87
                                                 $3,953      $4,038
```

4-17 $(A/P, 26\%, 12) = [0.26(1.26)^{12}]/[1.26^{12} - 1]$

```
          = 0.27732; (A/F,26%,12) = 0.27732 - 0.26 = 0.01732
CR:                                              Type A      Type B
$8,500(0.27732)-$850(0.01732) =                  $2,342
$11,000(0.27732)-$850(0.01732) =                             $3,036
Annual Operation & Maintenance                    2,550       2,125
                                                 $4,892      $5,161
```

4-18
```
CR: $25,000(A/P,9%,10)-$5,000(A/F,9%,10)
=   $25,000(0.15582)-$5,000(0.06582) =           $3,566
Annual Operation & Maintenance                    4,050
Extra Annual Income Tax                           1,260
                                                 $8,876
```
Plan B is preferable to Plan C by an annual cost of $8,876 - $8,477 = $399.

```
     Year      Plan B      Plan C      (C - B)
       0      -$15,000    -$25,000    -$10,000
     1-10      -6,140      -5,310        +830
      10          0        +5,000       +5,000
    Totals   -$76,400    -$73,100      +$3,300
```

EUAW(C-B) = -$10,000(0.15582) + $5,000(0.06582) + $830 = -$399. Thus Plan C entails an annual cost of $399 more than Plan B with interest at 9%.

27

PEE Solutions Manual   Chapter 4

4-19
		Plan I	Plan II
CR:			
$390,000(0.09)		$35,100	
$490,000(0.09)			$44,100
Annual Maintenance		12,000	4,000
Refurbishing Costs:			
$50,000(A/F,9%,15) = $50,000(0.03406) =		1,703	
$100,000(A/F,9%,10) = $100,000(0.06582) =		6,582	
$175,000(A/F,9%,25) = $175,000(0.01181) =			2,067
$250,000(A/F,9%,20) = $250,000(0.01955) =			4,888
		$55,385	$55,055

These two plans of development differ in perpetual annual cost by only $330.

4-20
The changes required to Problem 4-19 include only the provision for replacement of buildings and facilities for Plan I. Calculation of the replacement cost involves an additional $75,000 for buildings every 60 years and an additional $100,000 for facilities every 30 years. The equivalent uniform annual cost of these changes in assumptions amount to $75,000(A/F,9%,60) + $100,000(A/F,9%,30) = $75,000(0.00051) + $100,000(0.00734) = $38 + $734 = **$772**. Thus the total EUAC for Plan I becomes $55,385 + $772 = **$56,157**. Even with these very substantial increases in assumed future costs, the time value of money impact changes the final comparative EUAC's by only 1.4%.

4-21
	Type P	Type Q
CR:		
$60,000(A/P,20%,12) = $60,000(0.22526)	$13,516	
$85,000(A/P,20%,20)-$3,000(A/F,20%,20)		
= $85,000(0.20536)-$3,000(0.00536) =		$17,440
Relining Costs:		
$18,000[(P/F,20%,4)+(P/F,20%,8)](A/P,20%,12)		
= $18,000(0.4823 + 0.2326)(0.22526) =	2,899	
$12,000(P/F,20%,10)(A/P,20%,20)		
= $12,000(0.1615)(0.20536) =		398
Annual Operation & Maintenance:		
$2,600+$150(A/G,20%,12) = $2,600+$150(3.484) =	3,123	
$1,200+$50(A/G,20%,20) = $1,200+$50(4.464) =		1,423
	$19,538	$19,261

28

PEE Solutions Manual   Chapter 4

4-22
CR: Type P   Type Q

$60,000(A/P,20%,10) = $60,000(0.23852) =$ $14,311
$85,000(A/P,20%,10)-$3,000(A/F,20%,10)
= $85,000(0.23852)-$3,000(0.03852) =$ $20,159
Relining Costs:
$18,000[(P/F,20%,4)+(P/F,20%,8)](A/P,20%,10)
= $18,000(0.4823 + 0.2326)(0.23852) =$ 3,069
Annual Operation & Maintenance:
$2,600+$150(A/G,20%,10) = $2,600+$150(3.074) =$ 3,061
$1,200+$50(A/G,20%,10) = $1,200+$50(3.074) =$ 1,354
                                                             $20,441 $21,513
The economic choice is reversed under these new conditions.

4-23
CR: Machine A Machine B
$10,000(A/P,8%,6)-$1,000(A/F,8%,6)
= $10,000(0.21632)-$1,000(0.13632) =$ $2,027
$13,800(0.21632)-$1,000(0.13632) =$ $2,849
Annual Operation & Maintenance:
$800+$80(A/G,8%,6) = $800+$80(2.276) =$ 982
$200+$100(2.276) =$ 428
Extra Annual Income Tax = $120-$8(2.276) =$ 102
                                                             $3,009 $3,379

4-24
CR: Machine A Machine B
$10,000(A/P,16%,6)-$1,000(A/F,16%,6)
= $10,000(0.27139)-$1,000(0.11139) =$ $2,603
$13,800(0.27139)-$1,000(0.11139) =$ $3,634
Annual Operation & Maintenance:
$800+$80(A/G,16%,6) = $800+$80(2.073) =$ 966
$200+$100(2.073) =$ 407
                                                             $3,569 $4,041

4-25
No treatment: EUAC = $770(A/P,7%,20) = $770(0.09439) = $72.68

Treatment:
EUAC = [$770 + $8.5{(P/F,7%,10) + (P/F,7%,20)}](A/P,7%,30)
= [$770 + $8.5(0.5083 + 0.2584)](0.08059) = $62.58

This annual saving of $10.10 does not seem like very much. But, when it
is multiplied my the many thousands of wood poles a utility owns, it
amounts to a very substantial savings indeed.

PEE Solutions Manual   Chapter 4

4-26
Annual labor savings = 1,000(2)($12.5) = $25,000
Gradient on above = 1,000(2)($0.35) = $700
Maintenance & energy = $5,200
Property taxes and insurance = $42,000(0.03) = $1,260
Net annual savings = $18,540 + (G = $700)

(a) Analysis before income taxes

    NAW = -$42,000(A/P,20%,10) + $18,540 + $700(A/G,20%,10)
        = -$42,000(0.23852) + $18540 + $700(3.074) = <u>+$10,674</u>

(b) Analysis after income taxes

    NAW = -$42,000(A/P,10%,10) + $12,804 + $420(A/G,10%,10)
        = -$42,000(0.16275) + $12,804 + $420(3.725) = <u>+$7,533</u>

4-27

		Machine V	Machine W
CR:			
$12,000(A/P,11%,12) = $12,000(0.15403) =		$1,848	
$28,000(A/P,11%,18)-$6,400(A/F,11%,18)			
= $28,000(0.12984)-$6,000(0.01984) =			$3,517
Annual Operation & Maintenance		10,200	7,000
Extra Annual Income Tax			1,200
		$12,048	$11,717

4-28

		Method A	Method B
CR:			
$8,400(A/P,12%,6)-$1,200(A/F,12%,6)			
= $8,400(0.24323)-$1,200(0.12323) =		$1,895	
$5,600(0.24323)-$2,000(0.12323) =			$1,116
Annual Operation & Maintenance		660	1,500
Extra Annual Income Tax		96	
		$2,651	$2,616

4-29

		Method A	Method B
CR:			
$8,400(A/P,25%,6)-$1,200(A/F,25%,6)			
= $8,400(0.33882)-$1,200(0.08882) =		$2,740	
$5,600(0.33882)-$2,000(0.08882) =			$1,720
Annual Operation & Maintenance		660	1,500
		$3,400	$3,220

PEE Solutions Manual

CHAPTER 5

Present Worth

5-1

	Plan C	Plan D
Investment costs:	$500,000	$800,000
$200,000(A/F,7%,20)/0.07		
= $200,000(0.02439)/0.07 =		
$30,000(P/F,7%,30) = $30,000(0.1314) =	69,686	3,942
Annual disbursements:		
$30,000/0.07-$10,000(P/A,7%,20)		
= $428,571-$10,000(10.594) =	322,631	
$10,000/0.07 =		142,857
	$892,317	$946,799

5-2

	Gas	Electric
Investment costs:	$6,000	$8,000
-$1,000(P/F10%,5) = -$1,000(0.6209) =		- 621
Annual disbursements:		
$1,100(P/A,10%,5) = $1,100(3.791) =	4,170	
$780(3.791) =		2,957
	$10,170	$10,336

5-3

	Plan I	Plan II
Investment costs:	$50,000	$30,000
$25,000(P/F,9%,9) = $25,000(0.4604) =	11,510	
-$10,000(P/F,9%,18) = -$10,000(0.2120) =	- 2,120	
$30,000(P/F,9%,6) = $30,000(0.5963) =		17,889
$20,000(P/F,9%,12) = $20,000(0.3555) =		7,110
Annual disbursements:		
$11,000(P/A,9%,9) = $11,000(5.995) =	65,945	
$18,000[(P/A,9%,18)-(P/A,9%,9)]		
= $18,000(8.756-5.995) =	49,698	
$8,000(P/A,9%,6) = $8,000(4.486) =		35,888
$16,000[(P/A,9%,12)-(P/A,9%,6)]		
= $16,000(7.161-4.486) =		42,800
$25,000[(P/A,9%,18)-(P/A,9%,12)]		
= $25,000(8.756-7.161) =		39,875
	$175,033	$173,562

5-4

Two years already have elapsed. Thus we can assume that the first two $1,000 payments already have been made. The company can afford to pay the PW of the remaining 15 years' payments:

PW = $2,000(P/A,16%,15) + $3,000(P/A,16%,10)
   - $4,000(P/A,16%,2) = $2,000(5.575)
   + $3,000(4.833) - $4,000(1.605) = $19,229

PEE Solutions Manual   Chapter 5

5-5
The subsidy is the difference between the $2 million loan and the present worth of the $50,000 annual repayments discounted at 7% interest.
Subsidy = $2,000,000 - $50,000(P/A,7%,40)
        = $2,000,000 - $50,000(13.332) = $1,333,400

5-6
The maximum price that the XYZ Company can afford to pay for the property is the PW of the future royalties foregone and selling price of the land.

PW = $2(20,000)(P/A,14%,15) - $1(20,000)(P/A,14%,5)
   + $30,000(P/F,14%,15) = $40,000(6.142)
   - $20,000(3.433) + $30,000(0.1401) = $181,220

5-7
Since John Doe will have $2,000/yr additional expenses and an i* of 11%, the PW of the future earnings to him is:

PW = $38,000(P/A,11%,15) - $20,000(P/A,11%,5)
   + $30,000(P/F,11%,15) = $38,000(7.191)
   - $20,000(3.696) + $30,000(0.2090) = $205,610

Doe and XYZ Company are pretty far apart in their views of the value of this property. Some irreducible factors Doe should consider are his declining health, his need for cash, alternative uses of the cash, the liklihood of renewal of the contract, the future of the XYZ Company, his relationship with his heirs, etc.

5-8
	Plan I	Plan II
Investment costs:	$150,000	$250,000
$75,000(A/F,9%,30)/0.09		
= $75,000(0.00734)/0.09 =	6,117	
$120,000(A/F,9%,50)/0.09		
= $120,000(0.00123)/0.09 =		1,640
Annual disbursements:		
$7,000/0.09+$3,000(P/A,9%,10)		
= $7,000/0.09+$3,000(6.418) =	97,032	
$4,000/0.09 =		44,444
	$253,149	$296,084

PEE Solutions Manual   Chapter 5

5-9   Type A   Type B
   Investment costs: $135,000 $100,000
   $15,000[(P/F,6%,20)+P/F,6%,40)]
   = $15,000(0.3118+0.0972) =   6,135
   $20,000(0.3118+0.0972) =   8,180
   $80,000(P/F,6%,30) = $80,000(0.1741) =   13,928
   Annual disbursements:
   $3,000(P/A,6%,60)+$60(P/G,6%,60)
   = $3,000(16.161)+$60(239.043) =   62,826
   $4,000(16.161)+$80(239.043) =   83,767
                                                  $203,961 $205,875

5-10
   Investment costs: Model H   Model K
   $12,000+$9,500[(P/F,20%,3)+(P/F,20%,6)
   +(P/F,20%,9)]-$2,500(P/F,20%,12)
   = $12,000+$9,500(0.5787+0.3349+0.1938)
   -$2,500(0.1122) = $22,240
   $16,000+$13,500[(P/F,20%,4)+(P/F,20%,8)]
   -$2,500(P/F,20%,12) = $16,000
   +$13,500(0.4823+0.2326)-$2,500(0.1122) = $25,371
   Annual Disbursements:
   [$5,000+$300(A/G,20%,3)](P/A,20%,12)
   = [$5,000+$300(0.879)](4.439) =   23,366
   [$3,800+$200(A/G,20%,4)](P/A,20%,12)
   = [$3,800+$200(1.274)](4.439) =   17,999
                                                   $45,606 $43,370

5-11
   Investment costs: Unit A   Unit B
   $40,000[1+(P/F,14%,8)+(P/F,14%,16)
   +(P/F,14%,24)+(P/F,14%,32)]
   = $40,000(1+0.3506+0.1229+0.0431+0.0151) = $61,268
   $50,000+$45,000[(P/F,14%,10)+(P/F,14%,20)
   +(P/F,14%,30)]-$5,000(P/F,14%,40)
   = $50,000+$45,000(0.2697+0.0728
   +0.0196)-$5,000(0.0053) = $66,268
   Annual disbursements:
   $1,800(P/A,14%,40) = $1,800(7.105) =   12,789
   $1,640(7.105) =   11,652
                                                   $74,057 $77,920

PEE Solutions Manual   Chapter 5

5-12
Investment costs: Unit X   Unit Y
$16,000-$2,000(P/F,20%,10)
= $16,000-$2,000(0.1615) = $15,677
$24,000-$3,000(0.1615) = $23,516
Annual disbursements:
$9,000(P/A,20%,10)+$30(P/G,20%,10)
= $9,000(4.192)+$30(12.887) =   38,115
= $7,520(4.192)+$12(12.887) = $31,678
                                                           $53,792 $55,194

5-13
                                                           Plan I   Plan II
Investment costs: $390,000 $490,000
$50,000(A/F,9%,15)/0.09
= $50,000(0.03406)/0.09 =   18,922
$100,000(A/F,9%,10)/0.09
= $100,000(0.06582)/0.09 =   73,133
175,000(A/F,9%,25)/0.09
= $175,000(0.01181)/0.09 =   22,964
$250,000(A/F,9%,20)/0.09
= $250,000(0.01955)/0.09 =   54,306
Annual disbursements:
$12,000/0.09 =   133,333
$4,000/0.09 =   44,444
                                                           $615,388 $611,714

5-14
                                                           Model H   Model K
Investment costs: $12,000 $16,000
$9,500(A/F,20%,3)(P/A,20%,9)-$2,500(P/F,20%,12)
= $9,500(0.27473)(4.031)-$2,500(0.1122) =   10,240
$13,500(A/F,20%,4)(P/A,20%,8)-$2,500(P/F,20%,12)
= $13,500(0.18629)(3.837)-$2,500(0.1122) =   9,369
[$5,000+$300(A/G,20%,3)](P/A,20%,12)
= [$5,000+$300(0.879)](4.439) =   23,366
[$3,800+$200(A/G,20%,4)](P/A,20%,12)
= [$3,800+$200(1.274)](4.439) =   17,999
                                                           $45,606 $43,368

5-15
                                                           Type X   Type Y
Investment costs: $360,000 $440,000
$180,000[(P/F,9%,10)+(P/F,9%,20)]
= $180,000(0.4224+0.1784) =   108,144
$240,000(P/F,9%,15) = $240,000(0.2745) =   65,880
Annual upkeep:
$34,000(P/A,9%,30) = $34,000(10.274) =   349,316
$18,000(10.274) =   184,932
                                                           $817,460 $690,812

34

PEE Solutions Manual   Chapter 5

5-16   System A   System B
    Investment costs: $85,000 $105,000
    $24,000(P/F,9%,10) = $24,000(0.4224) =   10,138
    Annual disbursements:
    $8,000(P/A,9%,20)+$500(P/G,9%,20)
    = $8,000(9.129)+$500(61.777) =   103,921
    $6,000(9.129)+$350(61.777) =   76,396
                                                           $199,059 $181,396

5-17 Present lining:
    ($4,700+$3,000)[(P/F,20%,1)+(P/F,20%,3)+(P/F,20%,5)+(P/F,20%,7)
    +(P/F,20%,9)+(P/F,20%,11)+(P/F,20%,13)]
    = $7,700(0.8333+0.5787+0.4019+0.2791+0.1938+0.1346+0.0935)
    = $7,700(2.5149) = $19,365
    Proposed lining:
    ($10,000+$3,000)[(P/F,20%,1)+(P/F,20%,5)+(P/F,20%,9)+(P/F,20%,13)]
    = $13,000(0.8333+0.4019+0.1938+0.0935)
    = $13,000(1.5225) = $19,793

5-18
    With the shorter life, the relinings planned for year 13 will not be
made in either case. This changes the present worths found in Problem
5-17 to:

    Present lining: $19,365 - $7,700(P/F,20%,13)
                    = $19,365 - $7,700(0.0935) = $18,645
    Proposed lining: $19,793 - $13,000(0.0935) = $18,578

Whereas in Problem 5-17 the present method was preferred by a PW of $428,
with the prospect of a shorter life, the proposed new method is preferred
by a PW of $67. Considering the lack of certainty with respect to the
remaining life of the tank, the final decision is likely to be based on
factors which have not been reduced to money values.

5-19
    Investment costs: Structure C   Structure D
    $260,000+$220,000(P/F,9%,15)
    -$40,000(P/F,9%,30) = $260,000
    +$220,000(0.2745) - $40,000(0.0753) = $317,378
    $500,000 - $50,000(P/F,9%,30)
    = $500,000 - $50,000(0.0753) = $496,240
    Annual costs: $480,000(P/A,9%,30)
    = $480,000(10.274) =   493,152
    $260,000(10.274) =   267,124
                                                   $810,530 $763,364

35

PEE Solutions Manual   Chapter 5

5-20
	Plan X	Plan Y
Investment costs:	$1,700,000	$2,600,000
$1,700,000[(P/F,10%,20)+(P/F,10%,40)]		
= $1,700,000(0.1486+0.0221) =	290,190	
$2,390,000(P/F,10%,30)-$210,000(P/F,10%,60)		
= $2,390,000(0.0573)-$210,000(0.0033) =		136,254
Annual disbursememts:		
$260,000(P/A,10%,60) = $260,000(9.967) =	2,591,420	
$199,733(9.967) =		1,990,739
	$4,581,610	$4,726,993

5-21
NPW = -$170,000 + $10,000(P/F,15%,15) + $18,000(P/A,15%,15)
      + $500(P/G,15%,15) - $800(P/F,15%,9)(P/G,15%,6)
      - $6,500(P/A,15%,15) - $200(P/G,15%,15)
    = -$170,000 + $10,000(0.1229) + $11,500(5.847) + $300(26.693)
      - $800(0.2843)(7.937) = **-$95,328**

5-22

Yr.	Trojan	Giant	Giant-Trojan	(P/F,15%,n)	PW
0	-$80,000	-$120,000	-$40,000	1.0000	-$40,000
1	- 16,000	- 10,000	+ 6,000	0.8696	+ 5,218
2	- 27,600	- 10,800	+ 16,800	0.7561	+ 12,702
3	- 19,200	- 19,600	- 400	0.6575	- 263
4	- 92,800	- 12,400	+ 80,400	0.5718	+ 45,973
5	- 16,000	- 13,200	+ 2,800	0.4972	+ 1,392
6	- 27,600	-122,000	- 94,400	0.4323	- 40,809
7	- 19,200	- 10,000	+ 9,200	0.3759	+ 3,458
8	- 92,800	- 10,800	+ 82,000	0.3269	+ 26,806
9	- 16,000	- 19,600	- 3,600	0.2843	- 1,023
10	- 27,600	- 12,400	+ 15,200	0.2472	+ 3,757
11	- 19,200	- 13,200	+ 6,000	0.2149	+ 1,289
12	- 12,800	- 2,000	+ 10,800	0.1869	+ 2,019
	-$466,800	-$376,000	+$90,800		+$20,519

Interpretation: By subtracting the cash flow year by year for the Trojan from the Giant, and computing the present worth of that difference, the result is a positive value of $20,519. In other words, the extra investment in the Giant is recovered with an excess PW of $20,519 when interest is 15%. The PW of the Trojan cash flow series is -$255,799, that for the Giant is -$235,280.

PEE Solutions Manual   Chapter 5

5-23   2 Manuals   Automatic
    Investment costs: $73,000 $120,000
    -$10,000(P/F,8%,20) = -$10,000(0.2145) =   - 2,145
    -$8,500(0.2145) =   - 1,823
    Overhauls: $6,000(A/F,8%,5)(P/A,8%,15)
    = $6,000(0.17046)(8.559) =   8,754   8,754
    Replacements: $27,500(P/F,8%,10) = $27,500(0.4632)=   12,738
    Annual disbursements:
    $58,000(P/A,8%,15) = $58,000(9.818) =   569,444
    $45,980(9.818) =   451,432
                                                            $649,053 $604,551

5-24
    Semiannual interest = $10,000(0.05/2) = $250. Price now to yield a
nominal 7%/annum, 3.5% effective semiannually:

    PW = $250(P/A,3.5%,40)+$10,000(P/F,3.5%,40)
       = $250(21.355)+$10,000(0.2526) = $7,865

5-25
    The investor can afford to pay the PW of the net receipts:

    PW = $3,540(P/A,18%,15)+$25,000(P/F,18%,15)
       = $3,540(5.092)+$25,000(0.0835) = $20,113

5-26
    Present worth of the cash flows is: $17,100(P/A,12%,15)
    + $150,000(P/F,12%,15) = $17,100(6.811) + $150,000(0.1827)
    = $143,873. This is maximum offer the investor should make.
    If the offer is not accepted, invest the money elsewhere at 12%
    or more.

5-27
    Semi-annual interest payment is $100,000(0.06/2) = $3,000. The
current sales price at 4.5% semiannually for 17 periods will be:

    PW = $3,000(P/A,4.5%,17) + $100,000(P/F,4.5%,17)
       = $3,000(11.707) + $100,000(0.4732) = $82,441

5-28

The scholarship will reach $10,000 at the end of year 26. One way to set this up is as follows:

PW = $5,000/0.08 + $200(P/G,8%,25) + $5,000(P/F,8%,25)/0.08
   = $5,000/0.08 + $200(87.804) + $5,000(0.1460)/0.08
   = $62,500 + $17,561 + $9,125 = $89,186

5-29

The scholarship will reach the $10,000 level at the end of year 26 with the last award given at the end of year 30.

PW = $5,000(P/A,8%,30)+$200(P/G,8%,25)+$5,000(P/A,8%,5)(P/F,8%,25)
   = $5,000(11.258) + $200(87.804) + $5,000(3.993)(0.1460)
   = $56,290 + $17,561 + $2,915 = $76,766

5-30

The scholarship will reach the $10,000 level at the end of year 26. At the end of year 25, the amount remaining in the endowment fund will be:

F_{25} = $80,000(F/P,8%,25) - $5,000(F/A,8%,25)
      - $200(A/G,8%,25)(F/A,8%,25)
      = $80,000(6.8485) - $5,000(73.106) - $200(8.225)(73.106)
      = $547,880 - $365,530 - $120,259 = $62,091

The remaining period of the $10,000 scholarship may be found by interpolating on n for the value of (A/P,8%,n):

(A/P,8%,n) = $10,000/$62,091 = 0.16105

n is between 8 and 9. Thus, the last full $10,000 will be paid at the end of the 8th year. The 9th year scholarship to just exhaust the endowment will be:

[$62,091(F/P,8%,8) - $10,000(F/A,8%,8)](F/P,8%,1) =
[$62,091(1.8509) - $10,000(10.637)](1.08) = $9,239

Thus the full term of the scholarship will be 33 or 34 years depending upon how the final $9,239 scholarship is interpreted.

5-31

Semiannual interest payment on bond is $10,000(0.08/2) = $400
P = $400(P/A,i%/2,30) + $10,000(P/F,i%/2,30)
@6%: P = $400(19.600) + $10,000(0.4120) = $11,960
@7%: P = $400(18.392) + $10,000(0.3563) = $10,920
@9%: P = $400(16.289) + $10,000(0.2670) = $ 9,186
@10%: P = $400(15.372) + $10,000(0.2314) = $ 8,463

PEE Solutions Manual   Chapter 5

5-32
NPW = - $60,000 + $8,000(P/A,i%,10) + $50,000(P/F,i%,10)
@6%;   NPW = -$60,000 + $8,000(7.360) + $50,000(0.5584) = $26,800
@8%;   NPW = -$60,000 + $8,000(6.710) + $50,000(0.4632) = $16,840
@10%; NPW = -$60,000 + $8,000(6.144) + $50,000(0.3855) = $ 8,427
@12%; NPW = -$60,000 + $8,000(5.650) + $50,000(0.3220) = $ 1,300
@15%; NPW = -$60,000 + $8,000(5.019) + $50,000(0.2472) = -$7,488

PEE Solutions Manual

CHAPTER 6

Internal Rate of Return

General Notes

The rate of return that makes the equivalent uniform cash flow of two projects (cash flows) equal is also the prospective rate of return on the extra investment of the project requiring the larger initial negative cash flow. It is obvious that the determination of an unknown rate of return on an investment can be computed using either the present worth method or the equivalent uniform annual cash flow method. There is only one thing that must be carefully watched: the present worth method requires that the same service for the same length of time be compared. Therefore, whenever the competing projects have different lives, it may be easier to use the EUAC method to determine the unknown interest rate, at least by hand calculation.

Linear interpolation between two interest rates will introduce some error in the answer. It is worth noting that the error will be in one direction if the PW method is used and in the opposite direction if the EUAC method is used. Thus, if the method used is opposite to the one used in this solution manual, the reader may find a difference in the answers in the second decimal place. In general, since most cash flows in typical problems are estimates of the future, the authors ordinarily advise against computing answers to more than one decimal place in percent. In this solution manual, however, many of the solutions have been obtained by the use of a microcomputer. Consequently, many answers are given to two or even three decimal places in percent for the benefit of the instructor.

In previous chapters the authors frequently have omitted algebraic signs by identifying the cash flows as either receipts or disbursements. In this chapter and in the solution manual the algebraic signs have been indicated. The authors strongly advise students and instructors to use the proper signs in all solutions for several reasons. First, computers have come into common usage in most companies, universities and many homes. Computer programs generally require that all equations include the proper signs in order to carry out the instructions. Secondly, some students use special hand held calculators that have been programmed to solve engineering economy and financial problems. These calculators also require that proper signs be used. Finally, the convention of using the negative sign to indicate a disbursement and the positive sign to indicate a receipt is just good logic and their consistent use will assist students in understanding the process of economic analysis.

PEE Solutions Manual   Chapter 6

6-1

NPW = 0 = -$100,000 +$9,500(P/A,i%,5) +$500(P/G,i%,5)
            +$107,500(P/F,i%,5)
At i = 11%,
NPW = -$100,000 +$9,500(3.696) +$500(6.624) +$107,500(0.5935)
    = +$2,225
At i = 12%,
NPW = -$100,000 +$9,500(3.605) +$500(6.390) +$107,500(0.5674)
    = -$1,559
By interpolation, i = 11% + 1%($2,225/$3,784) = __11.6%__

6-2

(a) NPW = 0 = -$9,200 +$300(P/A,i%,16) +$10,000(P/F,i%,16)
    At i = 3.5%, NPW = -$9,200 +$300(12.094) +$10,000(0.5767)
                    = +$195.20
    At i = 4%,   NPW = -$9,200 +$300(11.652) +$10,000(0.5339)
                    = -$365.40
    By interpolation, i = 3.5% + 0.5($195.20/$560.60) = __3.674%__
    Nominal annual i = 2(3.674) = __7.348%__

(b) Effective annual i = $(1.03674)^2 -1$ = __7.48%__

6-3

Yr	Method A	Method B	(B - A)
0	-$40,000	-$55,000	-$15,000
1	- 12,000	- 9,000	+ 3,000
2	- 14,000	- 10,000	+ 4,000
3	- 16,000	- 11,000	+ 5,000
4	- 16,000	- 11,000	+ 5,000
5	- 16,000	- 11,000	+ 5,000
5	0	+ 5,000	+ 5,000

NPW = 0 = -$15,000 +$3,000(P/F,i%,1) +$4,000(P/F,i%,2)
        +$5,000(P/A,i%,3)(P/F,i%,2) +$5,000(P/F,i%,5)
At i = 18%, NPW = -$15,000 +$3,000(0.8475) +$4,000(0.7182)
            +$5,000(2.174)(0.7182) +$5,000(0.4371) = +$408
At i = 20%, NPW = -$15,000 +$3,000(0.8333) +$4,000(0.6944)
            +$5,000(2.106)(0.6944) +$5,000(0.4019)
          = -$401
By interpolation, __i = 19.0%__; Invest in B.

41

6-4
NPW = 0 = +$9,300,000 −$540,000(P/A,i%,40)
       −$10,000,000(P/F,i%,40)
At i = 5.5%, NPW = +$9,300,000 −$540,000(16.046)
       −$10,000,000(0.1175) = +$539,840
At i = 6%, NPW = +$9,300,000 −$540,000(15.046)
       −$10,000,000(0.0972) = −$203,160
By interpolation, i = **5.863%**; Nominal annual i = 2(5.863%) = **11.726%**

6-5
NPW = 0 = −$40,000 +($8,400 − $2,400)(P/A,i%,15)
       −($500 −$200)(P/G,i%,15) +$4,000(P/F,i%,15)
At i = 7%, NPW = −$40,000 +$6,000(9.108) −$300(52.446)
       +$4,000(0.3624) = +$363.8
At i = 8%, NPW = −$40,000 +$6,000(8.559) −$300(47.886)
       +$4,000(0.3152) = −$1,751
By interpolation, i = **7.17%**

6-6
Since the structure in Plan I has a 30-year life and that in Plan II has a 50-year life, the best solution is to equate the EUAC of Plan I to the EUAC of Plan II at unknown interest rate, i. If i is equal to or greater than i*, select Plan II.
       −$150,000 i −$75,000(A/F,i%,30) −$7,000 −$3,000(P/A,i%,10) i
       = −$250,000 i −$120,000(A/F,i%,50) −$4,000
0 = −$100,000 i +$75,000(A/F,i%,30) − $120,000(A/F,i%,50)
       +$3,000 +$3,000(P/A,i%,10) i
At 4½%, −$100,000(0.045) +$75,000(0.01639) −$120,000(0.00560)
       +$3,000 +$3,000(7.913)0.045 = +$125.5
At 5%, −$100,000(0.05) +$75,000(0.01505) −$120,000(0.00478)
       +$3,000 +$3,000(7.722)0.05 = −$286.6
By interpolation, i = **4.65%**

6-7
Equate PW of 60 years of service at unknown i:
Type A: NPW = −$120,000 −$15,000(1 +(P/F,i%,20)
       +(P/F,i%,40)) −$3,000(P/A,i%,60) −$60(P/G,i%,60)
Type B: NPW = −$80,000(1 + (P/F,i%,30)) −$20,000(1 +(P/F,i%,20)
       +(P/F,i%,40)) −$4,000(P/A,i%,60) −$80(P/G,i%,60)

	i = 6%	i = 7%
NPW Type A	−$203,961	−$193,141
NPW Type B	− 205,875	− 188,033
(A−B)	+ 1,914	− 5,108

i = 6% +1%($1,914/$7,022) = **6.27%**

6-8

Yr	Machine K	Machine J	(J-K)
0	-$30,000	-$50,000	-$20,000
1	- 8,000	-6,000-600	+ 1,400
2	- 8,500	-6,300-680	+ 1,520
'			
'	G= -$500	G= -$380	G= +$120
12			
12	0	+$14,000	+$14,000

NPW = 0 = -$20,000 +$1,400(P/A,i%,12) +$120(P/G,i%,12)
      +$14,000(P/F,i%,12)
Try 8% and 9%: i = 8.18%; Since i* = 12%, Select Machine K.

6-9

NPW = 0 = -$18,000 +$1,000(P/A,i%,2) +$5,000(P/A,i%,8)(P/F,i%,2)
      +$2,000(P/A,i%,5)(P/F,i%,10)
At i = 16%, NPW = -$18,000 +$1,000(1.605) +$5,000(4.344)(0.7432)
      +$2,000(3.274)(0.2267) = +$2,591.94
At i = 18%, NPW = -$18,000 +$1,000(1.566) +$5,000(4.078)(0.7182)
      +$2,000(3.127)(0.1911) = -$594.76
By interpolation, i = 17.27%; greater than i*. Buy patent.

6-10

Yr	Gas	Electric	(E-G)
0	-$6,000	-$8,000	-$2,000
1 - 5	- 1,100	- 780	+ 320
5	0	+ 1,000	+ 1,000

NPW = 0 = -$2,000 +$320(P/A,i%,5) +$1,000(P/F,i%,5)
At i = 7%, NPW = +$25.00
At i = 8%, NPW = -$41.64
By interpolation, i = 7.38%

6-11

Yr	Project Y	Project Z	(Z-Y)
0	-$100,000	-$140,000	-$40,000
1 - 20	+ 17,000	+ 19,900	+ 2,900

Project Y: NPW = 0 = -$100,000 +$17,000(P/A,i%,20)
            (P/A,i%,20) = 5.8824; i = 16.15%
Project Z: NPW = 0 = -$140,000 +$19,900(P/A,i%,20)
            (P/A,i%,20) = 7.0352; i = 12.98%
i on extra investment in Z over Y:
            NPW = 0 = -$40,000 +$2,900(P/A,i%,20)
            (P/A,i%,20) = 13.7931; i = 3.83%; Select Y.

6-12

Yr	Toltec	Mandan	(T-M)
0	-$8,000	-$6,400	-$1,600
1 - 10	- 3,232	- 3,540	+ 308

NPW = 0 = -$1,600 + $308(P/A,i%,10)
(P/A),i%,10) = ($1,600/$308) = 5.1948
By interpolation, i = **14.1%**

6-13

NPW = 0 = +$19,200 -$400 -$2,250(P/A,i%,20)
 -$20,000(P/F,i%,20)
Try i = 12%; NPW = -$79.25
Try i = 13%; NPW = $1,257.75
By interpolation, i = **12.059%**

6-14

Yr	Project X	Project Y	(Y-X)
0	-$440,000	-$575,000	-$135,000
1 - 25	+ 57,800	+ 65,500	+ 7,700
25	+ 50,000	+ 50,000	0

Project X: NPW = -$440 +$57.8(P/A,i%,25) +$50(P/F,i%,25)
i = **12.5%**, greater than i*. Project X is acceptable.
Rate of return on extra investment in Project Y = i;
NPW = -$135 +$7.7(P/A,i%,25)
 (P/A,i%,25) = (135/7.7) = 17.5325
By interpolation, i = **2.9%**, less than i*. Reject Project Y.

6-15

NPW = 0 = -$215,000 +$40,000(P/F,i%,10) +($73,000-$26,000)
 (P/A,i%,10); A/P = $47,000/$215,000 = 0.219. Thus:
At 20%, NPW = -$215 +$40(0.1615) +$47(4.192) = -$11.516
At 18%, NPW = -$215 +$40(0.1911) +$47(4.494) = +$3.862
Thus i = **18.5%** Accept the proposal.

6-16

NPW = 0 = -$215,000 +$40,000(P/F,i%,10) +$32,000(P/A,i%,10)
 +$1,500(P/G,i%,10)
At 12%, NPW = -$215 +$40(0.3220) +$32(5.650) +$1.5(20.254)
 = +$9.061
At 13%, NPW = -$215 +$40(0.2946) +$32(5.426) +$1.5(19.080)
 = -$0.964
i = **12.9%**, which is less than i*. Reject the proposal.

PEE Solutions Manual   Chapter 6

6-17
Yr	K-400	E-450	(E-K)
0	-$12,000	-$24,000	-$12,000
1 - 4	- 3,700	- 2,800	+ 900
4	- 11,000	0	+ 11,000
5 - 8	- 3,700	- 2,800	+ 900
8	+ 1,000	+ 2,000	+ 1,000

NPW = 0 = -$12,000 +$900(P/A,i%,8) +$11,000(P/F,i%,4)
          +$1,000(P/F,i%,8)
At i = 11%, NPW = +$311.0
At i = 12%, NPW = -$134.4
i = <u>11.7%</u>, greater than i*; Select E-450.

6-18
Month	Purchase (P)	Extra Income Tax	Lease (L)	(P-L)
0	-$7,000		-$200	-$6,800
1 - 35	0		- 200	+ 200
12		-$324		- 324
24		- 324		- 324
36		- 324		- 324
36	+ 1,000			+ 1,000

(a) Before income tax:
    NPW = 0 = -$6,800 +$200(P/A,i%,35) +$1,000(P/F,i%,36)
    i = <u>0.827%</u> by computer. Effective annual rate = <u>10.39%</u>
(b) After income tax:
    NPW = 0 = -$6,800 +$200(P/A,i%,35) +$1,000(P/F,i%,36)
          - $324((P/F,i%,12) + (P/F,i%,24) +(P/F,i%,36))
    By computer, i = <u>0.306%</u>; Effective annual rate = <u>3.73%</u>
(c) Leasing Company:
    NPW = 0 = -$6,200 +$200(P/A,i%,35) +$1,400(P/F,i%,36)
    i = <u>1.53%</u>; Effective annual rate i = <u>19.98%</u>

45

6-19

Yr	Buy	Make	(M-B)
0	0	-$50,000	-$50,000
1	-$20,000	- 9,600	+ 10,400
2	- 22,000	- 11,000	+ 11,000
⋮	G = - 2,000	G = - 1,400	G = + 600
5			
5	0	10,000	+ 10,000

NPW = 0 = -$50,000 +$10,400(P/A,i%,5) +$600(P/G,i%,5)
           +$10,000(P/F,i%,5)
At i = 9%, NPW = -$50 +$10.4(3.890) +$0.6(7.111) +$10(0.6499)
              = +$1.222
At i = 10%, NPW = -$50 +$10.4(3.791)+ $0.6(6.862)+ $10(0.6209)
              = -$0.247
By interpolation, i = **9.83%**
Since i* = 9% after taxes, this is an acceptable proposal.

6-20

For an i* of 17%, Plan 7 is still the one to select. It is necessary to calculate the after-tax rate of return on its increment of investment over Plan 5; this turns out to be 17.4%. (Plan 6 is unacceptable as compared to Plan 5 because its incremental rate of return over Plan 5 is less than 17%.)

For an i* of 12%, Plan 7 continues to be the one to select. The incremental rate of return of Plan 8 over Plan 7 is only 10.0%; subsequent incremental rates are negative.

Of course no plan is acceptable with an after-tax i* of 20% because all of the rates of return in line H are less than 20%.

PEE Solutions Manual   Chapter 6

6-21

		Onondaga	Oneida	Cayuga	Tuscarora	Seneca	Mohawk
A	Investment	$54,000	$60,500	$72,000	$77,400	$91,800	$108,000
B	Annual positive cash flow	20,800	21,600	26,500	27,800	32,700	36,800
C	Rate of return on total investment	35.0%	31.8%	33.1%	32.0%	31.7%	29.9%
D	Compare with		Onondaga	Onondaga	Cayuga	Cayuga	Seneca
E	Increment of investment		6,500	18,000	5,400	19,800	16,200
F	Increment of annual cash flow		800	5,700	1,300	6,200	4,100
G	Rate of return on increment of investment		neg.	27.0%	17.4%	26.6%	19.0%
H	Select	Onondaga		Cayuga	Cayuga	Seneca	Seneca

6-22

Refer to the table in Problem 6-21. Using an i* of 30%, Onondaga is acceptable. From that table, Oneida and Cayuga fail. Since all of the increments beyond Cayuga yield less than 30%, Onondaga would be the choice.

6-23

	Proposals:	A	B	C	D	E	F
A.	Investment	$2,000	$8,000	$16,000	$18,000	$24,000	$40,000
B.	Annual Savings	480	2,400	5,200	5,650	7,950	11,850
C.	Rate of return on total investment	20.7%	27.3%	30.2%	28.9%	30.9%	26.9%
D.	Compare with:		A	B	C	C	E
E.	Incremental invest.		$6,000	$8,000	$2,000	$8,000	$16,000
F.	Incremental return		1,920	2,800	450	2,300	3,900
G.	Incremental RoR		29.6%	33.0%	18.3%	25.9%	20.6%
H.	Select		B	C	C	E	E-F

Choice will be E at an i* of 25%, F at an i* of up to 20.6%.

PEE Solutions Manual   Chapter 6

6-24
This problem has been set up as an example of an Annual Worth formulation.

Plan A:
NAW = 0 = $176,000 −$55,000 −$200,000i −$400,000(A/P,i%,40)
i = 20.1%   Reject Plan A because prospective rate of return before income taxes is less than the 25% standard.

Plan B:
NAW = 0 = $260,000 −$70,000 −$200,000i −$560,000(A/P,i%,40)
i = 25.0%   Plan B is acceptable but must defend itself against the proposals that call for larger investments.

Plan C:
NAW = 0 = $380,000 −$100,000 −$200,000i −$800,000(A/P,i%,40)
i = 28.0%
To compare Plan C with Plan B:
NAW = 0 = $90,000 −$240,000(A/P,i%,40)
i = 37.5%   Plan C is preferable to Plan B. Its return on total investment and its return on the incremental investment both exceed the 25% standard.

Plan D:
NAW = 0 = $467,000 −$125,000 −$200,000i −$1,080,000(A/P,i%,40)
i = 26.8%
To compare Plan D with C:
NAW = 0 = $62,000 −$280,000(A/P,i%,40)
i = 22.1%   Plan D should be rejected because its incremental rate of return over Plan C is less than the 25% standard.

Plan E:
NAW = 0 = $598,000 −$150,000 −$200,000i −$1,520,000(A/P,i%,40)
i = 26.0%
To compare Plan E with Plan C:
NAW = 0 = $168,000 −$720,000(A/P,i%,40)
i = 23.3%   Plan E should be rejected because its incremental rate of return over Plan C is less than the 25% standard.
Select Plan C.

6-25
NPW = 0 = −$200,000 −$5,000(P/A,i%,5) +$352,000(P/F,i%,5)
At i = 9%, NPW = −$200 −$5(3.890) +$352(0.6499) = +$9.315
At i = 10%, NPW = −$200 −$5(3.791) +$352(0.6209) = −$0.398
i = 9.96%

6-26
NPW = 0 = −$200,000 − $5,000(P/A,i%,5) +$384,000(P/F,i%,5)
At i = 12%, NPW = −$200 − $5(3.605) +$384(0.5674) = −$0.143
i is just under 12%

48

6-27
NPW = 0 = -$3,750 +$187.50(P/A,i%,16) +$5,000(P/F,i%,16)
i = 6.27%
Nominal annual i = 2(6.27%) = 12.54%

Effective annual i = $(1.0627)^2$ -1 = 19.93%

6-28
If the Jones build on the lot:
NPW = 0 = -$25,000 -$300(P/A,i%,4) +$38,000(P/F,i%,4)
i = 10%; This is an attractive investment for the Jones.
If they sell the lot:
NPW = 0 = -$25,000 -$300(P/A,i%,4) +$33,040(P/F,i%,4)
i = 6.14%; This is considerably less than the 8.5% that the tax-free bonds are now paying. Irreducibles will play a major role in the final decision.

6-29
NPW = 0 = -$350,000 +($42,000 -$16,400)(P/A,i%,10)
      +$300,000(P/F,i%,10)
At i = 6%, NPW = -$350 +$25.6(7.360) +$300(0.5584) = +$5.936
At i = 7%, NPW = -$350 +$25.6(7.024) +$300(0.5083) = -$17.696
    i = 6.2%

6-30
NPW in 1982 = 0 = -$5,300 +$159(P/A,i%,3) +$200(P/A,i%,5)(P/F,i%,3)
     +$8,500(P/F,i%,8)
At i = 8%, NPW = +$336.22
At i = 9%, NPW = -$30.65; Thus i = 8.92%

6-31
Dealer's Proposal:
NPW = 0 = +$7,425 -$268(P/A,i%,36) -$75((P/F,i%,12) + (P/F,i%,24))
At i = 1.5%; (P/A,1.5%,36) = 27.6607
NPW = +$7,425 -$268(27.6607) -$75(0.8364 + 0.6995) = -$101.01
At i = 2%; (P/A,2%,36) = 25.4888
NPW = +$7,425 -$268(25.4888) -$75(0.7885 + 0.6217) = +$488.24
By interpolation, i = 1.586% (1.583% by computer)
Effective annual i = $(1.01586)^{12}$ -1 = 20.78% (20.74%)

Bank's proposal:
NPW = 0 = +$7,500 -$249(P/A,i%,36)
(P/A,i%,36) = ($7,500/$249) = 30.1205
i = 0.9975% (by computer)
Effective annual i = 1.009975^{12} -1 = 12.65%

6-32

Dealer's financing: APR = 1.9%, i = 1.9%/12 = 0.15833%;
   (A/P,0.15833%,36) = 0.028599
   Monthly payment = $9,600(0.028599) = $274.55

Credit Union financing: APR = 6.5%; i = 6.5%/12 = 0.54167%;
   (A/P,0.54167%,36) = 0.030649
   Monthly payment = $8,600(0.030649) = $263.58

It is clear that taking the cash rebate and financing at the higher rate is more economical.

We may find the Credit Union interest rate that makes the two offers equal by formulating:
$8,600(A/P,i%,36) = $274.55
(A/P,i%,36) = $274.55/$8,600 = 0.03192
from which i = 0.7723% and APR = 12(0.7723%) = 9.27%   At any APR interest rate below 9.27%, alternative financing is preferable to dealer financing.

PEE Solutions Manual

CHAPTER 7

Measures Involving Costs, Benefits and Effectiveness

General Notes

The problems of Chapter 7 are designed to introduce analysis by Benefit-Cost Ratio as a method for decision making in public projects. They help to bring out the fundamental fact that a great many of the benefit-cost analyses that have been used by government agencies have been incorrect because the incremental benefit - incremental cost approach has not been used. They also help to emphasize the fact that the interpretation of the benefit-cost ratio is frequently in error. The choice of the project that gives the highest ratio may or may not be best; an optimal choice depends on a number of other factors in the situation and on the alternatives available.

These ideas will become more clear if the student studies Chapter 16, Some Aspects of Economy Studies for Government Activities. The teacher may be able to stimulate students to think in terms of optimal allocation of resources and the effects of using erroneous benefit-cost methods.

7-1
(Solutions to 7-1 and 7-2 illustrate slightly different methods of handling gradient series.) (000) omitted from all money amounts.

For Location H
Investment $110
PW of maintenance costs = $35(P/A,7%,20) = $35(10.594) = 371
PW of road user costs = $300(P/A,7%,20) -$90(P/A,7%,10)
 +$10(P/G,7%,10) = $300(10.594) -$90(7.024)
 +$10(27.716) = <u>2,823</u>
 <u>$3,304</u>

For Location J; Investment minus residual value =
 $700 -$300(P/F,7%,20) = $700 -$300(0.2584) = $ 622
PW of maintenance costs =$21(P/A,7%,20) =$21(10.594) = 223
PW of road user costs = $225(P/A,7%,20) -$67.5(P/A,7%,10)
 +$7.5(P/G,7%,10) = $225(10.594) -$67.5(7.024)
 +$7.5(27.716) = <u>2,117</u>
 <u>$2,962</u>

For Location K; Investment minus residual value =
 $1,300 -$550(0.2584) = $1,158
PW of maintenance costs = $17(10.594) = 180
PW of road user costs = $195(10.594) -$58.5(7.024)
 +$6.5(27.716) = <u>1,835</u>
 <u>$3,173</u>

PEE Solutions Manual   Chapter 7

7-2
(000) omitted from money amounts.
For Location H
  CR = $110(A/P,7%,20) = $110(0.09439) = $10.4
  Annual maintenance costs =   35.0
  EUA road user costs = $300 -[$90(P/A,7%,9)
    -$10(P/G,7%,9)](A/P,7%,20) = $300 -[$90(6.515)
    -$10(23.140)](0.09439) =   266.5
                                                                    $311.9
For Location J
  CR = $700(A/P,7%,20) -$300(A/F,7%,20) =
    $700(0.09439) -$300(0.02439) = $58.8
  Annual maintenance costs =   21.0
  EUA road user costs = $225 -[$67.5(6.515)
    -$7.5(23.140)](0.09439) =   199.9
                                                                    $279.1

For Location K
  CR = $1,300(0.09439) -$550(0.02439) = $109.3
  Annual maintenance costs =   17.0
  EUA road user costs = $195 -[$58.5(6.515)
    -$6.5(23.140)](0.09439) =   173.2
                                                                    $299.5

7-3
For CI over NFC, an extra investment of $2,900,000 causes annual net favorable consequences of ($430,000 -$105,000) -$35,000 = $340,000.
(A/P,i%,50) = $340/$2,900 = 0.11724;
by interpolation, i= 11.7%
For D&R over CI, an extra investment of $2,400,000 causes annual net favorable consequences of ($105,000 -$55,000) -($28,000 +$10,000) -($40,000 -$35,000) = $7,000.
It is clear that the <u>prospective rate of return is negative</u> for a project with a 50-year life and zero terminal salvage value. It would take 343 years for an initial investment of $2,400,000 to be recovered by a $7,000 year benefit, even if i = 0%. Even if the project were assumed to have perpetual life, the prospective i would be only 0.3%.

7-4
Annual highway costs for M: (000) omitted.
  CR = $3,000(A/P,8%,20) -$1,800(A/F,8%,20)
    = $3,000(0.10185) -$1,800(0.02185) = $266.22
  Operation & Maintenance   120.00
                                                                   $386.22
Annual highway costs for N:
  CR = $5,000(0.10185) -$3,000(0.02185) = $443.70
  Operation & Maintenance =   90.00
                                                                   $533.70

B/C = ($880 -$660)/($533.70 -$386.22) = <u>1.49</u>

52

PEE Solutions Manual   Chapter 7

7-5
Annual highway costs are the same as in Problem 7-4, i.e.,
(000) omitted, $386.22 for M and $533.70 for N.
B = ($880 -$660) +($20 -$17)(A/G,8%,20) = $220 +$3(7.037) = $241.1
B/C = $241.1/($533.70 -$386.22) = <u>1.63</u>
Route N is preferred to M.

7-6
Government annual highway costs are (000 omitted):
    Route M $386.22; Route N $533.70 (See Problem 7-4).
Annual road user costs (000 omitted):
    Route M: $880 + $15(A/G,8%,20) = $880 +$15(7.037)
              = $985.6
    Route N: $660 +$11.5(7.037) = $740.9
B/C = ($985.6 -$740.9)/($533.70 -$386.22) = <u>1.66</u>

7-7
CI vs NFC. (000) omitted from all money amounts.
    C = $400(A/P,9%,30) +$30 = $400(0.09734) +$30 = $68.9
    B = ($300 -$180) + ($20 -$12)(A/G,9%,30)
      = $120 +$8(8.666) = $189.3; B/C = <u>2.75</u>
D&R vs CI.
    C = $1,600(0.09734) +$65 = $220.7
    B = ($180 -$30) +($12 -$5)(8.666) -$20 +$8 = $198.7
        B/C = <u>0.90</u>
CI + D&R vs NFC.
    C = $2,000(0.09734) +$95 = $289.7
    B = ($300 -$30) +($20 -$5)(8.666) -$20 +$8 = $388.0
        B/C = <u>1.34</u>
    Addition of D&R does not provide benefits in excess of costs.
Thus choose CI only.

7-8
NFC. (000) omitted from all money amounts.
PW = $300(P/A,9%,30) +$20(P/G,9%,30) = $300(10.274) +$20(89.028)
      = <u>$4,863</u>
CI. PW = $400 +$30(10.274) +$180(10.274) +$12(89.028)
      = <u>$3,626</u>
CI & D&R. PW = $2,000 +($95 +$30 +$20 -$8)(10.274) +$5(89.028)
      = <u>$3,853</u>
Channel Improvement alone has the lowest present worth of costs.

53

PEE Solutions Manual   Chapter 7

7-9
   CI vs NFC. (000) omitted from all money amounts.
      C = $400(A/P,5%,30) +$30 = $400(0.06505) +$30 = $56.0;
      B = ($300 -$180) +($20 -$12)(A/G,5%,30) = $120 +$8(10.969)
        = $207.8; B/C = **3.71**
   D&R vs CI.
      C = $1,600(0.06505) +$65 = 169.1
      B = ($180 -$30) +($12 -$5)(10.969) -$20 +$8 = $214.8
      B/C = **1.27**
   CI + D&R vs NFC.
      C = $2,000(0.06505) +$95 = $225.1
      B = ($300 -$30) +($20 -$5)(10.969) -$20 +$8 = $422.5
      B/C = **1.88**
   Using an i* of 5% rather than 9% results in justification of the additional investment in D&R.

7-10
   Considered by itself as an addition to the D&R project;
      C = ($100 -$15)(A/P,9%,30) +$5 = $85(0.09734) +$5
        = $13.3; B = $38; B/C = **2.86**
   However, the D&R project did not have a B/C ratio greater than 1. Therefore, D&R plus this modification should be considered together. D&R modified vs. CI:
      C = $1,685(0.09734) +$70 = $234.0;
      B = ($180 -$30) +($12 -$5)(8.666) -$20 +$8 +$38 = $236.7;
      B/C = **1.01**. This modification changes the D&R project from being unacceptable to barely acceptable. The overall B/C ratio for CI and modified D&R will be:
      C = $2,085(0.09734) +$80 = $283.0; B = ($300 -$30)
         +($20 -$5)(8.666) -$20 +$8 +$38 = $426.0;
      B/C = **1.51**

7-11
   Table 7-3 shows CR costs of $403 and $484 for alternatives B-3 and B-4. With 100% terminal salvage values these figures become $350 and $420 (7% of the respective investments). The increment of benefits of B-4 over B-3 continues to be $70 as shown in Table 7-4. The increment of costs now is ($420 +$40 -$350 -$30) = $80. The incremental B/C ratio for B-4 over B-3 therefore is $70 ÷ $80 = 0.88. It follows that B-3 is preferable to B-4 even with the assumption of 100% terminal salvage value.

PEE Solutions Manual   Chapter 7

7-12

Annual costs: n = 20   50   100
   (A/P,5%,n)   = 0.08024   0.05478   0.05038

Plan	EUAC
A:	$5,000,000(0.05038) +$60,000 = $311,900
B:	$311,900 +$1,000,000(0.05478) +$25,000 = $391,680
C:	$311,900 +$125,120 = $437,020
D:	$391,680 +$125,120 = $516,800
E:	$3,750,000(0.05038) +$50,000 = $238,925
F:	$238,925 +$125,120 = $364,045
G:	($800,000 −$300,000)(0.08024) +$300,000(0.05) +$70,000 = $125,120

Benefit/cost ratios for total investments:

	Total Benefits	Total EUAC	Ratio
A:	$580,000	$311,900	1.86
B:	760,000	391,680	1.94
C:	640,000	437,020	1.46
D:	820,000	516,800	1.59
E:	500,000	238,925	2.09
F:	590,000	364,045	1.62
G:	350,000	125,120	2.80

Incremental B/C ratio analysis:

Defender	Challenger	Increment of Benefits	Increment of costs	Ratio	Winner
No project	G	$350,000	$125,120	2.80	G
G	E	150,000	113,805	1.32	E
E	A	80,000	72,975	1.10	A
A	F	10,000	52,145	0.19	A
A	B	180,000	79,780	2.26	B
B	C	−120,000	45,340	neg.	B
B	D	60,000	125,120	0.48	B

Select Plan B, Willow dam, reservoir, and power plant.

PEE Solutions Manual   Chapter 7

7-13

```
         Annual costs       n = 20         n = 50          n = 100
         (A/P,9%,n) =       0.10955        0.09123         0.09002
Plan                     EUAC
  A   $5,000,000(0.09002) +$60,000 = $510,100
  B   $510,100 +$1,000,000(0.09123) +$25,000 = $626,330
  C   $510,100 +$151,775 = $661,875
  D   $626,330 +$151,775 = $778,105
  E   $3,750,000(0.09002) +$50,000 = $387,575
  F   $387,575 +$151,775 = $539,350
  G   ($800,000 -$300,000)(0.10955) + $300,000(0.09) +$70,000
      = $151,775
```

Benefit/Cost ratios for total investments

Plan	Total Benefits	Total EUAC	Ratio
A	$580,000	$510,100	1.14
B	760,000	626,330	1.21
C	640,000	661,875	0.97
D	820,000	778,105	1.05
E	500,000	387,575	1.29
F	590,000	539,350	1.09
G	350,000	151,775	2.31

Incremental B/C ratio analysis:

Defender	Challenger	Increment of Benefits	Increment of Costs	Ratio	Winner
No project	G	$350,000	$151,775	2.31	G
G	E	150,000	235,800	0.64	G
G	A	230,000	358,325	0.64	G
G	F	240,000	387,575	0.62	G
G	B	410,000	474,555	0.86	G
G	C	290,000	510,100	0.57	G
G	D	470,000	626,330	0.75	G

Select G, channel improvement alone.

7-14

CR + Annual Maintenance Costs:
Site A: $12,000,000(0.05601) +$200,000 = $872,120
Site B: $16,000,000(0.05601) +$240,000 = $1,136,160
CI: $4,000,000(0.07455) +$330,000 = $628,200
A+CI: $872,120 +$628,200 = $1,500,320
B+CI: $1,136,160 +$628,200 = $1,764,360

Arrange the five plans in increasing order of EUAC:

Plan	Annual Benefit	Annual Cost	B/C Ratio	Increment Analyzed	Increment of Benefit	Increment of Cost	B/C
CI	$1,300,000	$628,200	2.07	--	$1,300,000	$628,200	2.07
A	1,697,500	872,120	1.95	A-CI	397,500	243,500	1.63
		Eliminate CI					
B	2,007,500	1,136,160	1.77	B-A	310,000	264,040	1.17
		Eliminate A					
A+CI	2,245,000	1,500,320	1.50	A+CI-B	237,500	364,160	0.65
		Eliminate A+CI					
B+CI	2,395,000	1,764,360	1.36	B+CI-B	387,500	628,200	0.62
		Eliminate B+CI					

Choose Site B alone, because it is the largest investment that meets the criterion of a B/C ratio equal to or greater than 1 for each increment of investment.

Computation of the total equivalent uniform annual cost, including remaining flood damage, will lead to the same conclusion.

7-15

CR + Annual Maintenance Costs:
Site A: $18,000,000(A/P,8%,75) +$300,000 = $1,744,500
Site B: $24,000,000(0.08025) +$360,000 = $2,286,000
CI   : $6,000,000(0.09368) +$495,000 = $1,057,080
A + CI: $1,744,500 +$1,057,080 = $2,801,580
B + CI: $2,286,000 +$1,057,080 = $3,343,080

Arrange the five plans in increasing order of investment required.

Plan	Annual Benefits	Annual Cost	B/C Ratio	Increment Analyzed	Incr. of Benefits	Incr. of Costs	B/C
CI	$2,600,000	$1,057,080	2.46				
A	3,395,000	1,744,500	1.95	A-CI	$795,000	$687,420	1.16
		Eliminate CI					
B	4,015,000	2,286,000	1.76	B-A	620,000	541,500	1.14
		Eliminate A					
A+CI	4,490,000	2,801,580	1.60	A+CI-B	475,000	515,580	0.92
		Eliminate A + CI					
B+CI	4,790,000	3,343,080	1.43	B+CI-B	775,000	1,057,080	0.73
		Eliminate B + CI					

Choose Site B alone.

7-16
CR + Annual Maintenance Costs:
(A/P,8%,10) = 0.14903 ; (A/P,8%,50) = 0.08174
R-0: $100,000
R-1: $700,000(0.14903) +$70,000 = $174,321
R-2: $700,000(0.14903) +$500,000(0.08174) +$70,000 = $215,191
R-3: $600,000(0.14903) +$1,400,000(0.08174) +$60,000 = $263,854
T-1: $550,000(0.14903) +$3,000,000(0.08174) +$55,000 = $382,187
T-2: $500,000(0.14903) +$4,000,000(0.08174) +$50,000 = $451,475

| | Annual | | | | Incremental | | |
Proposal	Benefit	Cost	B/C	Comparison	Benefit	Cost	B/C
R-0		$100,000					
R-1	$120,000	174,321	0.69	R1-R0	$120,000	$74,321	1.61
R-2	160,000	215,191	0.74	R2-R1	40,000	40,870	0.98
R-3	220,000	263,854	0.83	R3-R1	100,000	89,533	1.12
T-1	290,000	382,187	0.76	T1-R3	70,000	118,333	0.59
T-2	340,000	451,475	0.75	T2-R3	120,000	187,621	0.64

In no case do project benefits exceed project costs. If one project must be chosen, however, choose R-3. The minimum total EUAC is $263,854 + $700,000 = $963,854.

7-17
CR + Annual Maintenance Costs:
(A/P,4%,10) = 0.12329 ; (A/P,4%,50) = 0.04655
R-0: $100,000
R-1: $700,000(0.12329) +$70,000 = $156,303
R-2: $700,000(0.12329) +$500,000(0.04655) +$70,000 = $179,578
R-3: $600,000(0.12329) +$1,400,000(0.04655) +$60,000 = $199,144
T-1: $550,000(0.12329) +$3,000,000(0.04655) +$55,000 = $262,460
T-2: $500,000(0.12329) +$4,000,000(0.04655) +$50,000 = $297,845

| | Annual | | | | Incremental | | |
Proposal	Benefit	Cost	B/C	Comparison	Benefit	Cost	B/C
R-0		$100,000					
R-1	$120,000	156,303	0.77				
R-2	160,000	179,578	0.89				
R-3	220,000	199,144	1.11				
T-1	290,000	262,460	1.10	T1-R3	$70,000	$63,316	1.11
T-2	340,000	297,845	1.14	T2-T1	50,000	35,385	1.41

With i* = 4% rather than 8% (Problem 7-16) Proposal T-2 is preferred and has an overall B/C ratio of 1.14.

PEE Solutions Manual   Chapter 7

7-18
CR + Annual Maintenance Costs:
(A/P,12%,10) = 0.17698 ; (A/P,12%,50) = 0.12042
R-0: $100,000
R-1: $700,000(0.17698) +$70,000 = $193,886
R-2: $700,000(0.17698) +$500,000(0.12042) +$70,000 = $254,096
R-3: $600,000(0.17698) +$1,400,000(0.12042) +$60,000 = $334,776
T-1: $550,000(0.17698) +$3,000,000(0.12042) +$55,000 = $513,599
T-2: $500,000(0.17698) +$4,000,000(0.12042) +$50,000 = $620,170

	Annual				Incremental		
Proposal	Benefit	Cost	B/C	Comparison	Benefit	Cost	B/C
R-0		$100,000					
R-1	$120,000	193,886	0.62	R1-R0	$120,000	$93,886	1.28
R-2	160,000	254,096	0.63	R2-R1	40,000	60,210	0.66
R-3	220,000	334,776	0.66	R3-R1	100,000	140,890	0.71
T-1	290,000	513,599	0.56	T1-R1	170,000	319,713	0.53
T-2	340,000	620,170	0.55	T2-R1	220,000	426,284	0.52

At i* = 12%, no proposal has a benefit that exceeds costs. If one proposal must be accepted it should be R-1 with a minimum total EUAC of $993,886.

7-19
Since bids A & B essentially afford the same protection, a choice between them may be based on minimum Present Worth of costs.
PW(A) = $500,000 +$100,000[(P/F,9%,5) +(P/F,9%,10)]
      + $60,000(P/A,9%,15) = $500,000 +$100,000(0.6499 +0.4224)
      +$60,000(8.061) = $1,090,890
PW(B) = $640,000 +$48,000(8.061) = $1,026,928
B is preferred to A. In comparing C to B, it must be remembered that C offers a 15% improvement in efficiency.
PW(C) = $760,000 +$36,800(8.061) = $1,056,645
The 15% improvement in efficiency can be obtained at an equivalent of an additional PW of $29,717.
Factors the community should consider in a final decision include possible future changes in requirements and the value of a cleaner river. It may want to go into further negotiation with bidders B and C.

PEE Solutions Manual   Chapter 7

7-20
PW(A) = $500,000 + $100,000[(P/F,4%,15) +(P/F,4%,10)]
       + $60,000(P/A,4%,15) = $500,000 +$100,000(0.8219+ 0.6756)
       + $60,000(11.118) = $1,316,830
PW(B) = $640,000 +$48,000(11.118) = $1,173,664
PW(C) = $760,000 +$36,800(11.118) = $1,169,142
In this case, Bid C clearly is preferred. The low interest rate encourages greater initial investment and, in this case, favors investment in a more efficient system.

7-21
The following calculations use the equivalent uniform annual cost method and the data from Table 7-2. The same answers will be obtained using the present worth method and the information from Table 7-1. (000) omitted in all the money amounts.

For J over H, B/C = $\frac{($266.5 +$35) -($199.9 +$21)}{$58.8 -$10.4} = \frac{$80.6}{$48.4} = \underline{1.67}$

For K over J, B/C = $\frac{($199.9 +$21) -($173.2 +$17)}{$109.3 -$58.8} = \frac{$30.7}{$50.5} = \underline{0.61}$

For K over H, B/C = $\frac{($266.5 +$35) -($173.2 +$17)}{$109.3 -$10.4} = \frac{$111.3}{$98.9} = \underline{1.13}$

7-22
(000) omitted in all money amounts.
For D&R over CI, B/C = $50/($157 +$38) = $50/$190 = <u>0.26</u>
For D&R over NFC: B = $480 -$55 = $425;
    C = $5,300(A/P,6%,50) +$40 +$38 = $5,300(0.06344) +$78
    = $414; B/C = $425/$414 = <u>1.03</u>

7-23
B/C ratios for the four schemes of classification are:
(000) omitted.
(a) ($2,200 -$600)/($600 +$1,400) = 0.80
(b) $2,200/($600 +$1,400 +$600)   = 0.85
(c) ($2,200 -$600 -$1,400)/($600) = 0.33
(d) ($2,200 -$1,400)/($600 +$600) = 0.67
The second classification, which classifies the $300,000 as a cost, gives the most favorable ratio. The third scheme, which classifies O&M expenses as a "disbenefit", gives the lowest ratio.

PEE Solutions Manual

CHAPTER 8

Some Relationships Between
Accounting and Engineering Economy

General Notes

Since all investments in depreciable assets affect in some way the taxable income of the organization and in turn the cash flow for income tax payments, the analyst must be able to estimate the effects of the proposal on the taxable income. To do that the analyst must estimate the depreciation to be charged against the proposed project. Thus it is necessary to understand the basic concepts of depreciation accounting.

Depreciation accounting is a process of allocating to each year's operating costs a portion of the original investment in depreciable assets. These charges represent a reduction in annual income on which income taxes must be paid. Engineering economy studies deal with the future consequences of an investment. Familiarity with the depreciation accounting rules and practices is necessary in order to be able to predict the effects of depreciation charge allocations on income taxes paid in the future.

It is important for the teacher to convey accounting concepts to students so they can always distinguish the significant difference between a written financial history of an organization and the prediction of the effects of a decision on the future financial health of the organization.

What can and cannot be done in connection with depreciation accounting is spelled out in great detail by the tax laws passed by the US Congress and by the governing bodies of other countries. In the USA, significant changes were made in the rules and regulations regarding depreciation in 1954, 1981, and 1986, with less significant changes made in between. Ordinarily, one would only study the current laws, but it is not uncommon, especially in a replacement economy study, to be dealing with an investment that was made five to fifteen years earlier and has been affected by two or more sets of depreciation regulations passed in different years. It is for this reason that all basic methods, straight-line, sum-of-years-digits, and declining-balance, are discussed and illustrated in this chapter. After the 1986 tax laws became fully implemented in 1988, depreciation rates were limited to a certain number of classes of assets and to specified percentages of write-off each year. These percentages were derived from various combinations of the basic methods discussed.

The chapter describes the problems of accounting and engineering economy in very general terms based on the 1986 Act, but the details of the Act and its applications are reserved for Chapter 9 and Appendix F. Appendix F will be revised as frequently as necessary to keep the information up-to-date relative to the latest tax laws.

8-1

(a) $P = \$24,750$; $S = 0$; $n = 10$
SOYD: $SYD = (10 \times 11)/2 = 55$
$D_{10} = \$24,750(1/55) = \underline{\$450}$

(b) DDB: $f = 2(1/10) = 0.20$
$BV_{10} = \$24,750(1 - 0.2)^{10} = \$24,750(0.8)^{10} = \$2,657.51$
$BV_9 = \$24,750(0.8)^9 = \$3,321.89$
$D_{10} = \$3,321.89 - \$2,657.51 = \underline{\$664.38}$

(c) Yes, there will be a loss on disposal of $\underline{\$2,657.51}$

8-2

(a) SOYD: $SYD = (40 \times 41)/2 = 820$
$BV_{10} = \$100,000 + \$500,000 - (\$500,000/820)(820 - 465) = \underline{\$383,537}$
$BV_{20} = \$100,000 + \$500,000 - (\$500,000/820)(820 - 210) = \underline{\$228,049}$

(b) DDB: $f = 2/40 = 0.05$
$BV_{10} = \$100,000 + \$500,000(1 - 0.05)^{10} = \underline{\$399,368}$
$BV_{20} = \$100,000 + \$500,000(1 - 0.05)^{20} = \underline{\$279,243}$

8-3

$P = \$26,000$; $S = \$2,000$; $n = 15$; $i = 10\%$

(a) SL: $D_4 = (\$26,000 - \$2,000)/15 = \underline{\$1,600}$
$BV_4 = \$26,000 - \$1,600(4) = \underline{\$19,600}$

(b) SOYD: $SYD = (15 \times 16)/2 = 120$
$D_4 = \$24,000(12/120) = \underline{\$2,400}$
$BV_4 = \$26,000 - \$200(120 - 66) = \underline{\$15,200}$

(c) DDB: $f = 2/15 = 0.13333$
$BV_4 = \$26,000(1 - 0.13333)^4 = \underline{\$14,671}$
$BV_3 = \$26,000(1 - 0.13333)^3 = \$16,927$
$D_4 = \$16,927 - \$14,671 = \underline{\$2,256}$

8-4

(a) SOYD: $SYD = (5 \times 6)/2 = 15$
$D_1 = \$35,000(5/15) = \underline{\$11,667}$

Profit = $\$9,000 - (\$11,667 - \$7,000) = \underline{\$4,333}$

(b) DDB: $f = 2.000/5 = 0.40$
$D_1 = \$35,000(0.40) = \underline{\$14,000}$

Profit = $\$9,000 - (\$14,000 - \$7,000) = \underline{\$2,000}$

PEE Solutions Manual   Chapter 8

8-5
The following cash flow table for Proposal A shows a before-tax rate of return of 19.2% and after-tax return of 10.6%

	A	B	C	D	E	F
1			Write-off		Influence	Cash flow
2		Cash flow	of initial	Influence	of income	after
3		before	outlay for	on	taxes on	income
4		income	Tax	taxable	cash flow	taxes
5	year	taxes	purposes	income	-0.40D	(B+E)
6	0	($ 110,000)				($ 110,000)
7	1	$ 38,000	($ 11,000)	$ 27,000	($ 10,800)	$ 27,200
8	2	$ 34,000	($ 11,000)	$ 23,000	($ 9,200)	$ 24,800
9	3	$ 30,000	($ 11,000)	$ 19,000	($ 7,600)	$ 22,400
10	4	$ 26,000	($ 11,000)	$ 15,000	($ 6,000)	$ 20,000
11	5	$ 22,000	($ 11,000)	$ 11,000	($ 4,400)	$ 17,600
12	6	$ 18,000	($ 11,000)	$ 7,000	($ 2,800)	$ 15,200
13	7	$ 14,000	($ 11,000)	$ 3,000	($ 1,200)	$ 12,800
14	8	$ 10,000	($ 11,000)	($ 1,000)	$ 400	$ 10,400
15	9	$ 6,000	($ 11,000)	($ 5,000)	$ 2,000	$ 8,000
16	10	$ 2,000	($ 11,000)	($ 9,000)	$ 3,600	$ 5,600
17	Sums	$ 90,000	($ 110,000)	$ 90,000	($ 36,000)	$ 54,000
18						
19		19.2%	= IRR Before Tax		IRR A/T =	10.6%
20						

PEE Solutions Manual   Chapter 8

8-5 (cont)

The following cash flow table for Proposal B, using the same headings, shows a before-tax rate of return of 11.9% and the after-tax return of 8.3%.

	A	B	C	D	E	F
1			Write-off		Influence	Cash flow
2		Cash flow	of initial	Influence	of income	after
3		before	outlay for	on	taxes on	income
4		income	Tax	taxable	cash flow	taxes
5	year	taxes	purposes	income	-0.40D	(A + D)
6	0	($110,000)				($110,000)
7	1	$5,000	($11,000)	($6,000)	$2,400	$7,400
8	2	$9,000	($11,000)	($2,000)	$800	$9,800
9	3	$13,000	($11,000)	$2,000	($800)	$12,200
10	4	$17,000	($11,000)	$6,000	($2,400)	$14,600
11	5	$21,000	($11,000)	$10,000	($4,000)	$17,000
12	6	$25,000	($11,000)	$14,000	($5,600)	$19,400
13	7	$29,000	($11,000)	$18,000	($7,200)	$21,800
14	8	$33,000	($11,000)	$22,000	($8,800)	$24,200
15	9	$37,000	($11,000)	$26,000	($10,400)	$26,600
16	10	$41,000	($11,000)	$30,000	($12,000)	$29,000
17	Sums	$120,000	($110,000)	$120,000	($48,000)	$72,000
18						
19			11.9% = IRR Before Tax		IRR A/T =	8.3%
20						

64

8-6
Annual book values and depreciation charges for Proposal A, using the double-declining-balance method of depreciation, are shown in the following table.

$f = 2.000/10 = 0.20$

	A	B	C	D	E	F	G
1		Book	Depreciation		Book	Depreciation	
2	Year	Value	Charge		Value	Charge	
3	0	$110,000			$110,000		
4	1	$88,000	$22,000		$90,000	$20,000	
5	2	$70,400	$17,600		$72,000	$18,000	
6	3	$56,320	$14,080		$56,000	$16,000	
7	4	$45,056	$11,264		$42,000	$14,000	
8	5	$36,045	$9,011		$30,000	$12,000	
9	6	$28,836	$7,209		$20,000	$10,000	
10	7	$23,069	$5,767		$12,000	$8,000	
11	8	$18,455	$4,614		$6,000	$6,000	
12	9	$14,764	$3,691		$2,000	$4,000	
13	10	$11,811	$2,953		$0	$2,000	
14							
15	Sum		$98,189			$110,000	
16							
17	Prob.8-6 A				Prob.8-6 B		

Loss on disposal with double-declining-balance method is $11,811.

PEE Solutions Manual   Chapter 8

8-7

(a) SOYD Method, Proposal A

	A	B	C	D	E	F
1			Write-off		Influence	Cash flow
2		Cash flow	of initial	Influence	of income	after
3		before	outlay for	on	taxes on	income
4		income	Tax	taxable	cash flow	taxes
5	year	taxes	purposes	income	−0.40D	(B+E)
6	0	($ 110,000)				($ 110,000)
7	1	$ 38,000	($ 20,000)	$ 18,000	($ 7,200)	$ 30,800
8	2	$ 34,000	($ 18,000)	$ 16,000	($ 6,400)	$ 27,600
9	3	$ 30,000	($ 16,000)	$ 14,000	($ 5,600)	$ 24,400
10	4	$ 26,000	($ 14,000)	$ 12,000	($ 4,800)	$ 21,200
11	5	$ 22,000	($ 12,000)	$ 10,000	($ 4,000)	$ 18,000
12	6	$ 18,000	($ 10,000)	$ 8,000	($ 3,200)	$ 14,800
13	7	$ 14,000	($ 8,000)	$ 6,000	($ 2,400)	$ 11,600
14	8	$ 10,000	($ 6,000)	$ 4,000	($ 1,600)	$ 8,400
15	9	$ 6,000	($ 4,000)	$ 2,000	($ 800)	$ 5,200
16	10	$ 2,000	($ 2,000)	$ 0	$ 0	$ 2,000
17	Sums	$ 90,000	($ 110,000)	$ 90,000	($ 36,000)	$ 54,000
18					IRR A/T =	11.8%
19						
20	Prob.8-7A					

8-7 (cont)
(b) SOYD Method, Proposal B

	A	B	C	D	E	F
1			Write-off		Influence	Cash flow
2		Cash flow	of initial	Influence	of income	after
3		before	outlay for	on	taxes on	income
4		income	Tax	taxable	cash flow	taxes
5	year	taxes	purposes	income	-0.40D	(A + D)
6	0	($ 110,000)				($ 110,000)
7	1	$ 5,000	($ 20,000)	($ 15,000)	$ 6,000	$ 11,000
8	2	$ 9,000	($ 18,000)	($ 9,000)	$ 3,600	$ 12,600
9	3	$ 13,000	($ 16,000)	($ 3,000)	$ 1,200	$ 14,200
10	4	$ 17,000	($ 14,000)	$ 3,000	($ 1,200)	$ 15,800
11	5	$ 21,000	($ 12,000)	$ 9,000	($ 3,600)	$ 17,400
12	6	$ 25,000	($ 10,000)	$ 15,000	($ 6,000)	$ 19,000
13	7	$ 29,000	($ 8,000)	$ 21,000	($ 8,400)	$ 20,600
14	8	$ 33,000	($ 6,000)	$ 27,000	($ 10,800)	$ 22,200
15	9	$ 37,000	($ 4,000)	$ 33,000	($ 13,200)	$ 23,800
16	10	$ 41,000	($ 2,000)	$ 39,000	($ 15,600)	$ 25,400
17	SUMS	$ 120,000	($ 110,000)	$ 120,000	($ 48,000)	$ 72,000
18						
19					IRR A/T =	8.9%
20	Prob. 8-7B					

(c) The SOYD method reduces the early cash flow for income taxes compared with the straight-line method, thus increasing the after-tax rates of return from 8.3% to 8.9% and from 10.6% to 11.8%. The total tax payments over the lives of the proposals will be the same. Only the timing of payments is different.

PEE Solutions Manual   Chapter 8

8-8
(a)

	A	B	C	D	E	F
1			Write-off		Influence	Cash flow
2		Cash flow	of initial	Influence	of income	after
3		before	outlay for	on	taxes on	income
4		income	Tax	taxable	cash flow	taxes
5	year	taxes	purposes	income	-0.40D	(A + D)
6	0	($ 45,000)				($ 45,000)
7	1	$ 9,000	($ 9,000)	$ 0	$ 0	$ 9,000
8	2	$ 12,000	($ 9,000)	$ 3,000	($ 1,200)	$ 10,800
9	3	$ 15,000	($ 9,000)	$ 6,000	($ 2,400)	$ 12,600
10	4	$ 18,000	($ 9,000)	$ 9,000	($ 3,600)	$ 14,400
11	5	$ 21,000	($ 9,000)	$ 12,000	($ 4,800)	$ 16,200
12	SUMS	$ 30,000	($ 45,000)	$ 30,000	($ 12,000)	$ 18,000
13						
14		17.0%	= IRR B/T		IRR A/T =	11.2%
15						

(b)

19	0	($ 45000)				($ 45000)
20	1	$ 20000	($ 9000)	$ 11000	($ 4400)	$ 15600
21	2	$ 17000	($ 9000)	$ 8000	($ 3200)	$ 13800
22	3	$ 14000	($ 9000)	$ 5000	($ 2000)	$ 12000
23	4	$ 11000	($ 9000)	$ 2000	($ 800)	$ 10200
24	5	$ 8000	($ 9000)	($ 1000)	$ 400	$ 8400
25	Sums	$ 25000	($ 45000)	$ 25000	($ 10000)	$ 15000
26						
27		20.1%	= IRR B/T		IRR A/T =	11.7%
28						

PEE Solutions Manual   Chapter 8

8-9
(a)

	A	B	C	D	E	F
1			Write-off		Influence	Cash flow
2		Cash flow	of initial	Influence	of income	after
3		before	outlay for	on	taxes on	income
4		income	Tax	taxable	cash flow	taxes
5	year	taxes	purposes	income	-0.40D	(A + D)
6	0	($ 45,000)				($ 45,000)
7	1	$ 9,000	($ 18,000)	($ 9,000)	$ 3,600	$ 12,600
8	2	$ 12,000	($ 10,800)	$ 1,200	($ 480)	$ 11,520
9	3	$ 15,000	($ 6,480)	$ 8,520	($ 3,408)	$ 11,592
10	4	$ 18,000	($ 4,860)	$ 13,140	($ 5,256)	$ 12,744
11	5	$ 21,000	($ 4,860)	$ 16,140	($ 6,456)	$ 14,544
12	SUMS	$ 30,000	($ 45,000)	$ 30,000	($ 12,000)	$ 18,000
13						
14			17.0%	= IRR B/T		IRR A/T =	12.0%
15						

(b)

18						
19	0	($ 45000)				($ 45000)
20	1	$ 20000	($ 18000)	$ 2000	($ 800)	$ 19200
21	2	$ 17000	($ 10800)	$ 6200	($ 2480)	$ 14520
22	3	$ 14000	($ 6480)	$ 7520	($ 3008)	$ 10992
23	4	$ 11000	($ 4860)	$ 6140	($ 2456)	$ 8544
24	5	$ 8000	($ 4860)	$ 3140	($ 1256)	$ 6744
25	Sums	$ 25000	($ 45000)	$ 25000	($ 10000)	$ 15000
26						
27			20.1%	= IRR B/T		IRR A/T =	12.9%
28						

PEE Solutions Manual   Chapter 8

8-10

	A	B	C	D	E	F	G
1		10%	10%		15%		
2		S.L. Dep.			S.L. Dep.		
3	Year	+Av.Int.	(A/P,i,n)	Error	+Av. Int.	(A/P,i,n)	Error
4	5	0.26	0.2638	0.0038	0.29	0.29832	0.00832
5	10	0.155	0.16275	0.00775	0.1825	0.19925	0.01675
6	15	0.12	0.13147	0.01147	0.14667	0.17102	0.02435
7	20	0.1025	0.11746	0.01496	0.12875	0.15976	0.03101
8	25	0.092	0.11017	0.01817	0.118	0.1547	0.0367
9	30	0.085	0.10608	0.02108	0.11083	0.1523	0.04147
10							
11							

70

PEE Solutions Manual   Chapter 8

8-11

P = $452,000; i = 20% before tax; S = $200,000; n = 40 years
(a) EUAC = -$452,000(A/P,20%,40) +$200,000(A/F,20%,40)
         = -$452,000(0.20014) +$200,000(0.00014) = **$90,434**

(b)

	A	B	C	D
1		S.L.	SOYD	
2	Year	Deprec.	Deprec.	(B-C)
3	0			
4	1	($ 11,000)	($ 21,333)	$ 10,333
5	2	($ 11,000)	($ 20,666)	$ 9,666
6	3	($ 11,000)	($ 19,999)	$ 8,999
7	4	($ 11,000)	($ 19,332)	$ 8,332
8	5	($ 11,000)	($ 18,665)	$ 7,665
9	6	($ 11,000)	($ 17,998)	$ 6,998
10	7	($ 11,000)	($ 17,331)	$ 6,331
11	8	($ 11,000)	($ 16,664)	$ 5,664
...	...			
26	23	($ 11,000)	($ 6,659)	($ 4,341)
27	24	($ 11,000)	($ 5,992)	($ 5,008)
28	25	($ 11,000)	($ 5,333)	($ 5,667)
29	26	($ 11,000)	($ 4,667)	($ 6,333)
30	27	($ 11,000)	($ 4,000)	($ 7,000)
31	28	($ 11,000)	($ 3,333)	($ 7,667)
32	29	($ 11,000)	($ 2,667)	($ 8,333)
33	30	($ 11,000)	($ 2,000)	($ 9,000)
34	31	($ 11,000)	($ 1,333)	($ 9,667)
35	32	($ 11,000)	($ 666)	($ 10,334)
36	Sums	($ 352,000)	($ 352,000)	($ 0)
37			NPV @ 20% =	$ 35,200

The rapid write-off by SOYD vs. SL would increase the true after-tax rate of return substantially. The NPV before tax, at i = 20%, is $35,200.

8-12

	A	B	C	D	E
1	8-12	P = $55,000	S = $0;	n= 10	
2		Year-By-Year Depreciation Charges			
3		Straight-	Sum-Of-	Double-	1986 Tax
4	Year	Line	Years Dig'	Dec'g. Bal.	Act
5	1	($5,500)	($10,000)	($11,000)	($15,714)
6	2	($5,500)	($9,000)	($8,800)	($11,224)
7	3	($5,500)	($8,000)	($7,040)	($8,017)
8	4	($5,500)	($7,000)	($5,632)	($5,727)
9	5	($5,500)	($6,000)	($4,506)	($4,773)
10	6	($5,500)	($5,000)	($3,604)	($4,772)
11	7	($5,500)	($4,000)	($2,306)	($4,772)
12	8	($5,500)	($3,000)	($1,845)	
13	9	($5,500)	($2,000)	($1,476)	
14	10	($5,500)	($1,000)	($1,181)	
15	Sums	($55,000)	($55,000)	($47,390)	-55000

The 1986 7-year class depreciation rates provide for much faster write-off of the investment than either of the other three methods. Therefore, its use will give a higher after-tax rate of return than would be attained with either of the others.

8-13

A firm may logically choose to use straight-line depreciation instead of the 1986 Tax Act depreciation schedule if:

(a) It has very little profit now and expects the profit to increase substantially in future years.

(b) It is now in one of the lower tax brackets and expects its profitability to push it into higher tax brackets in future years.

(c) It has reason to believe that the tax rates will be increased substantially in future years.

8-14

(a) $D_{10} = (\$30,000 - \$4,000)/25 = \underline{\$1,040}$

$BV_{10} = \$30,000 - \$1,040(10) = \underline{\$19,600}$

(b) SOYD $= 25(26/2) = 325$; $G = (\$30,000 - \$4,000)/325 = \$80$
$D_{10} = \$80(16) = \underline{\$1,280}$
$BV_{10} = \$30,000 - \$80(325 - 15(16/2)) = \underline{\$13,600}$

(c) $f = 2.00/25 = 0.08$
$BV_9 = \$30,000(1 - 0.08)^9 = \$14,165$
$BV_{10} = \$30,000(1 - 0.08)^{10} = \$13,032$
$D_{10} = BV_9 - BV_{10} = \underline{\$1,133}$

8-15

(a) $f = 1 - \sqrt[n]{S/P}$; $f = 1 - \sqrt[10]{0.02}$
$= 1 - 0.6762 = 0.3238$ or $\underline{32.38\%}$

(b) $f = 1 - \sqrt[10]{0.05} = \underline{25.89\%}$

(c) $f = 1 - \sqrt[10]{0.20} = \underline{14.87\%}$

8-16

(a)

Year A	C F for Purchase B	C F for Make C	Net C F (C-B) D
0	0	-$26,000	-$26,000
1	-$16,000	- 10,000	+ 6,000
2	- 18,500	- 12,000	+ 6,500
3	- 19,500	- 12,000	+ 7,500
4	- 14,500	- 9,000	+ 5,500
5	- 10,000	- 6,000	+ 4,000
6	- 6,000	- 3,000	+ 3,000
6	0	+ 2,000	+ 2,000
Totals	-$84,500	-$76,000	+ $8,500

Find i so that NPW = 0 = -$26,000 +$6,000(P/F,i%,1)
+$6,500(P/F,i%,2) +$7,500(P/F,i%,3) +$5,500(P/F,i%,4)
+$4,000(P/F,i%,5) +$5,000(P/F,i%,6)
$\underline{i = 9.34\%}$ by interpolation between 9% and 10%.

PEE Solutions Manual   Chapter 8

8-16 (cont)
(b)

year	Cash flow before income taxes	Write-off of initial outlay for Tax purposes	Influence on taxable income	Influence of income taxes on cash flow -0.40D	Cash flow after income taxes (B+C)
0	($ 26,000)				($ 26,000)
1	$ 6,000	($ 4,000)	$ 2,000	($ 800)	$ 5,200
2	$ 6,500	($ 4,000)	$ 2,500	($ 1,000)	$ 5,500
3	$ 7,500	($ 4,000)	$ 3,500	($ 1,400)	$ 6,100
4	$ 5,500	($ 4,000)	$ 1,500	($ 600)	$ 4,900
5	$ 4,000	($ 4,000)	$ 0	$ 0	$ 4,000
6	$ 3,000	($ 4,000)	$ 1,000	($ 400)	$ 4,600
6	$ 2,000				
				IRR A/T =	4.8%

8-17
SL: BV_{10} = $1,100,000 − $1,000,000(10/20) = $600,000
      Find i so that:
NPW = 0 = −$1,100,000 + $120,000(P/A,i%,10) + $700,000(P/F,i%,10)
      By interpolation: <u>i = 8.45%</u>

74

PEE Solutions Manual   Chapter 8

8-18

Year	Invest.	Depreciation straight line	1986 tax act	SL minus 86 Tax Act
0	($10,000)			
1		($1,667)	($4,000)	$2,333
2		($1,667)	($2,400)	$733
3		($1,667)	($1,440)	($227)
4		($1,667)	($1,080)	($587)
5		($1,667)	($1,080)	($587)
6		($1,665)		($1,665)

NPV @20% = $1,245

8-19
(a) SL: D = $150,000/15 = $10,000$
    BV = $200,000 − $10,000(10) = $100,000$
(b) SOYD: SYD = 15(16)/2 = 120; G = $150/120 = $1,250
    D = $1,250(6) = $7,500$
    BV = $200,000 − ($150,000 − $1,250(5)(6/2)) = $68,750$
(c) D.D.B.: f = 2.00/15 = 0.13333
    BV_9 = $50,000 + $150,000(1 − 0.13333)^9 = $91,377
    BV_{10} = $50,000 + $150,000(1 − 0.13333)^{10} = $85,860$
    D = $91,377 − $85,860 = $5,517$

8-20
(a) SL: D = ($25,000 − $4,000)/6 = $3,500$
    BV = $25,000 − $3,500(3) = $14,500$
(b) SOYD: SYD = 6(7/2) = 21; G = $21,000/21 = $1,000
    D = $1,000(4) = $4,000$
    BV = $25,000 − $1,000(21 − 3(4/2)) = $10,000$
(c) DDB: f = $2,000/6 = 0.33333

    BV(2) = $25,000(1 − 0.33333)^2 = $11,111
    BV(3) = $25,000(1 − 0.33333)^3 = $7,408$
    D = $11,111 − $7,408 = $3,703$

8-21

Year	(a) Straight Line Net C F Before Dep.	Deprecia- tion	Profit Before Income Tax	(b) Double Declin. Bal Deprecia- tion	Profit Bef. Income Tax
1	$ 16,000	($ 7,000)	$ 9,000	($ 14,000)	$ 2,000
2	$ 19,000	($ 9,000)	$ 10,000	($ 12,400)	$ 6,600
3	$ 19,000	($ 9,000)	$ 10,000	($ 7,440)	$ 11,560
4	$ 17,000	($ 9,000)	$ 8,000	($ 4,460)	$ 12,540
Sums	$ 71,000	($ 34,000)	$ 37,000	($ 38,300)	$ 32,700

BV(S.L.) = $45000 -$34,000 = $11,000
BV(D.D.B.) = $45,000 - $38,300 = $6,700

8-22

Year	Net C F Before Deprec.	Straight Line Deprec.	Profit Before Income Tax
1	+$16,000	$3,500	$12,500
2	+ 19,000	4,500	14,500
3	+ 19,000	4,500	14,500
4	+ 17,000	4,500	12,500
Totals	$71,000	$17,000	$54,000

8-23

(a) Low-Lift Platform Trucks:
EUAC = -6,000($6.50) = -$39,000
Forklift Truck:
EUAC = -$15,000(A/P,20%,10) -$2,000 -2,000($14.00)
= -$3,578 -$2,000 -$28,000 = -$33,578

(b) Actual charges to overhead account:
Low-lift: $39,000 + any depreciation on old platform trucks
Fork trucks: First 5 years = $15,000/5 + $2,000 + $28,000
= $33,000
For second 5 years = $2,000 + $28,000 = $30,000

(c) Direct labor hours of work on the products presumably will not be affected by the change in materials handling methods, however charges to the overhead account will be reduced by the acquisition of the forklift truck. Therefore the total overhead charges will be reduced and the current labor hour rate of $2.75 should be reduced accordingly.

8-24

	A	B	C	D	E	F	G
1		P = $10,000; S = $2,800; n = 8					
2							
3		S. L.	Book	SOYD	Book	1986 Tax	Book
4	Year	Deprec.	Value	Deprec.	Value	Act Depre.	Value
5	0		$10,000		$10,000		$10,000
6	1	($900)	$9,100	($1,600)	$8,400	($4,000)	$6,000
7	2	($900)	$8,200	($1,400)	$7,000	($2,400)	$3,600
8	3	($900)	$7,300	($1,200)	$5,800	($1,440)	$2,160
9	4	($900)	$6,400	($1,000)	$4,800	($1,080)	$1,080
10	5	($900)	$5,500	($800)	$4,000	($1,080)	$0
11	6	($900)	$4,600	($600)	$3,400		$0
12	7	($900)	$3,700	($400)	$3,000		$0
13	8	($900)	$2,800	($200)	$2,800		$0

PEE Solutions Manual   Chapter 8

8-25
(a)

	A	B	C	D	E	F
1			Write-off		Influence	Cash flow
2		Cash flow	of initial	Influence	of income	after
3		before	outlay for	on	taxes on	income
4		income	Tax	taxable	cash flow	taxes
5	year	taxes	purposes	income	-0.40D	(B+E)
6	0	($10,000)				($10,000)
7	1	$3,000	($900)	$2,100	($840)	$2,160
8	2	$3,000	($900)	$2,100	($840)	$2,160
9	3	$3,000	($900)	$2,100	($840)	$2,160
10	4	$3,000	($900)	$2,100	($840)	$2,160
11	5	$3,000	($900)	$2,100	($840)	$2,160
12	6	$3,000	($900)	$2,100	($840)	$2,160
13	7	$3,000	($900)	$2,100	($840)	$2,160
14	8	$3,000	($900)	$2,100	($840)	$4,960
15	8	$2,800				
16					IRR A/T =	16.7%
17						
18						
19	0	($10,000)				($10,000)
20	1	$3,000	($4,000)	($1,000)	$400	$3,400
21	2	$3,000	($2,400)	$600	($240)	$2,760
22	3	$3,000	($1,440)	$1,560	($624)	$2,376
23	4	$3,000	($1,080)	$1,920	($768)	$2,232
24	5	$3,000	($1,080)	$1,920	($768)	$2,232
25	6	$3,000		$3,000	($1,200)	$1,800
26	7	$3,000		$3,000	($1,200)	$1,800
27	8	$3,000		$5,800	($2,320)	$1,800
28	8	$2,800				$2,800
29					IRR A/T =	21.0%

(b) is at row 17-18.

78

PEE Solutions Manual   Chapter 8

8-26

	A	B	C	D	E
1	8-26	P(1) = $35,000; S(1) = $4,750, n(1) =			
2		P(2) = $50,000; S(2) = $6,000, n(2) = 10			
3		Cash Flow	S. L.		
4		Before	Deprec.		
5	Year	Taxes	Charge		
6	0	($ 35,000)			
7	1		($ 3,025)		
8	2		($ 3,025)		
9	3		($ 3,025)		
10	4	($ 50,000)	($ 3,025)		
11	5		($ 7,425)		
12	6		($ 7,425)		
13	7		($ 7,425)		
14	8		($ 7,425)		
15	9		($ 7,425)		
16	10		($ 7,425)		
17	10	$ 4,750			

PEE Solutions Manual   Chapter 8

8-27

	A	B	C	D	E	F	G
1		Capital	Annual	Write-off		Influence	Cash flow
2		Invest'ts	Cash flow	of initial	Influence	of income	after
3		And	before	outlay for	on	taxes on	income
4		Salvage	income	Tax	taxable	cash flow	taxes
5	year	Values	taxes	purposes	income	-0.40D	(B + C + F)
6	0	($ 35,000)					($ 35,000)
7	1		$ 15,500	($ 3,025)	$ 12,475	($ 4,990)	$ 10,510
8	2		$ 15,500	($ 3,025)	$ 12,475	($ 4,990)	$ 10,510
9	3		$ 15,500	($ 3,025)	$ 12,475	($ 4,990)	$ 10,510
10	4	($ 50,000)	$ 15,500	($ 3,025)	$ 12,475	($ 4,990)	($ 39,490)
11	5		$ 15,500	($ 7,425)	$ 8,075	($ 3,230)	$ 12,270
12	6		$ 15,500	($ 7,425)	$ 8,075	($ 3,230)	$ 12,270
13	7		$ 15,500	($ 7,425)	$ 8,075	($ 3,230)	$ 12,270
14	8		$ 15,500	($ 7,425)	$ 8,075	($ 3,230)	$ 12,270
15	9		$ 15,500	($ 7,425)	$ 8,075	($ 3,230)	$ 12,270
16	10		$ 15,500	($ 7,425)	$ 8,075	($ 3,230)	$ 40,620
17	10	$ 4,750					
18	10	$ 23,600				IRR A/T =	15.0%

8-28

	A	B	C	D	E	F
1						
2	Year	Investment	Rate	Write-off	Book Value	Percentage
3	0	$ 1.00000				
4	1		28.57143%	$ 0.28571	$ 0.71429	71.42857%
5	2			$ 0.20408	$ 0.51020	51.02041%
6	3			$ 0.14577	$ 0.36443	36.44315%
7	4			$ 0.10412	$ 0.26031	26.03082%
8	5		8.67694%	$ 0.08677	$ 0.17354	17.35388%
9	6		8.67694%	$ 0.08677	$ 0.08677	8.67694%
10	7		8.67694%	$ 0.08677	$ 0.00000	0.00000%
11						
12						
13						
14		Write-off in 4 years		$ 0.73969		

PEE Solutions Manual

CHAPTER 9

Estimating Income Tax Consequences of Certain Decisions

9-1

year	Cash flow before income taxes	Write-off of initial outlay for Tax purposes	Influence on taxable income	Influence of income taxes on cash flow −0.40 C	Cash flow after income taxes (A + D)
0	($ 126,000)				($ 126,000)
1	$ 34,000	($ 28,000)	$ 6,000	($ 2,400)	$ 31,600
2	$ 34,000	($ 24,500)	$ 9,500	($ 3,800)	$ 30,200
7	$ 34,000	($ 7,000)	$ 27,000	($ 10,800)	$ 23,200
8	$ 34,000	($ 3,500)	$ 30,500	($ 12,200)	$ 21,800
		$ 0			
	21.3%	=IRR B/T		IRR A/T=	14.6%

9-2

(Headings same as Problem 9-1)

0	($ 126,000)				($ 126,000)
1	$ 34,000	($ 15,750)	$ 18,250	($ 7,300)	$ 26,700
1 to 8	$ 34,000	($ 15,759)	$ 18,241	($ 7,296)	$ 26,704
	21.3%	=IRR B/T		IRR A/T=	13.5%

PEE Solutions Manual   Chapter 9

9-3
(Headings same as Problem 9-1)

0	($126,000)			$12,600	($113,400)
1	$34,000	($26,600)	$7,400	($2,960)	$31,040
2	$34,000	($23,275)	$10,725	($4,290)	$29,710
3	$34,000	($19,950)	$14,050	($5,620)	$28,380
:	:	:	:	:	:
8	$34,000	($3,325)	$30,675	($12,270)	$21,730
				IRR A/T =	17.6%

9-4
(Headings same as Problem 9-1)

0	($126,000)				($126,000)
1	$34,000	($36,000)	($2,000)	$800	$34,800
2	$34,000	($30,000)	$4,000	($1,600)	$32,400
3	$34,000	($24,000)	$10,000	($4,000)	$30,000
4	$34,000	($18,000)	$16,000	($6,400)	$27,600
5	$34,000	($12,000)	$22,000	($8,800)	$25,200
6	$34,000	($6,000)	$28,000	($11,200)	$22,800
7	$34,000		$34,000	($13,600)	$20,400
8	$34,000		$34,000	($13,600)	$20,400
				IRR A/T =	15.3%

PEE Solutions Manual   Chapter 9

9-5
(Headings same as Problem 9-1)

0	($126,000)			$12,600	($113,400)
1	$34,000	($34,200)	($200)	$80	$34,080
2	$34,000	($28,500)	$5,500	($2,200)	$31,800
3	$34,000	($22,800)	$11,200	($4,480)	$29,520
4	$34,000	($17,100)	$16,900	($6,760)	$27,240
5	$34,000	($11,400)	$22,600	($9,040)	$24,960
6	$34,000	($5,700)	$28,300	($11,320)	$22,680
7	$34,000		$34,000	($13,600)	$20,400
8	$34,000		$34,000	($13,600)	$20,400
				IRR A/T =	18.5%

9-6
(Headings same as Problem 9-1)

0	($126,000)				($126,000)
1	$34,000	($6,300)	$27,700	($11,080)	$22,920
2	$34,000	($6,300)	$27,700	($11,080)	$22,920
to	⋮	⋮	⋮	⋮	⋮
8	$34,000	($6,300)	$27,700	($11,080)	$22,920
9	$0	($6,300)	($6,300)	$2,520	$2,520
to	⋮	⋮	⋮	⋮	⋮
19	$0	($6,300)	($6,300)	$2,520	$2,520
20	$0	($6,300)	($6,300)	$2,520	$2,520
				IRR A/T =	10.8%

9-7
(Headings same as Problem 9-1)

0	($ 126,000)				($ 126,000)
1	$ 34,000	($ 50,400)	($ 16,400)	$ 6,560	$ 40,560
2	$ 34,000	($ 30,240)	$ 3,760	($ 1,504)	$ 32,496
3	$ 34,000	($ 18,144)	$ 15,856	($ 6,342)	$ 27,658
4	$ 34,000	($ 13,608)	$ 20,392	($ 8,157)	$ 25,843
5	$ 34,000	($ 13,608)	$ 20,392	($ 8,157)	$ 25,843
6	$ 34,000		$ 34,000	($ 13,600)	$ 20,400
7	$ 34,000		$ 34,000	($ 13,600)	$ 20,400
8	$ 34,000		$ 34,000	($ 13,600)	$ 20,400
				IRR A/T =	15.8%

9-8
(Headings same as Problem 9-1)

0	($ 110,000)	($ 110,000)	($ 110,000)	$ 66,000	($ 44,000)
1	$ 26,300		$ 26,300	($ 10,520)	$ 15,780
2	$ 26,300		$ 26,300	($ 10,520)	$ 15,780
3	$ 26,300		$ 26,300	($ 10,520)	$ 15,780
4	$ 26,300		$ 26,300	($ 10,520)	$ 15,780
5	$ 26,300		$ 26,300	($ 10,520)	$ 15,780
6	$ 26,300		$ 26,300	($ 10,520)	$ 15,780
7	$ 26,300		$ 26,300	($ 10,520)	$ 15,780
8	$ 26,300		$ 26,300	($ 10,520)	$ 15,780
9	$ 26,300		$ 26,300	($ 10,520)	$ 15,780
10	$ 26,300		$ 26,300	($ 10,520)	$ 15,780
				IRR A/T =	33.9%

PEE Solutions Manual　　　　　　　　　　　　　　　　　　　　　　　Chapter 9

9-9

Consider a $20,000 asset, the example used in the problem statement. The solution requires evaluating the difference in tax saving under the two proposals, the 7% ITC and the 10% ITC. This analysis assumes tax rates of 30% and 50%, lives of 10 and 20 years, and SL and SOYD Depreciation.

10-yr. life, SL Dep. — Tax Savings

Yr	7% ITC 30%TR	50%TR	10% ITC 30%TR	50%TR	Difference 7%-10% 30%TR	50%TR
0	$420	$700	$600	$1,000	-$180	-$300
1-10	600	1,000	540	900	+ 60	+ 100
Rate of Return using 7% ITC rather than 10%					31.1%	31.1%

20-yr. life, SL Dep. — Tax Savings

Yr	7% ITC 30%TR	50%TR	10% ITC 30%TR	50%TR	Difference 7%-10% 30%TR	50%TR
0	$420	$700	$600	$1,000	-$180	-$300
1-20	300	500	270	450	+ 30	+ 50
Rate of Return using 7% ITC rather than 10%					15.8%	15.8%

10-yr. life, SOYD Dep. — Tax Savings

Yr	7% ITC 30%TR	50%TR	10% ITC 30%TR	50%TR	Difference 7%-10% 30%TR	50%TR
0	$420	$700	$600	$1,000	-$180	-$300
1	1,091	1,818	982	1,636	+ 109	+ 182
2	982	1,636	884	1,472	+ 98	+ 164
3	873	1,455	786	1,308	+ 87	+ 146
.	.	.	.	.	.	.
10	109	182	98	164	+ 11	+ 18
Rate of Return using 7% ITC rather than 10%					48.5%	48.5%

In general, the 7% ITC without reduction in the depreciation base is preferable to the 10% ITC with a corresponding 10% reduction in the depreciation base. More liberal depreciation methods enhance this preference considerably. A preference for the 10% ITC as described may occur for extremely long lived assets and particularly if SL Depreciation is required. Buildings would fit in this category. This results from the fact that the break-even rate of return decreases as depreciable life increases. This can be seen from the rate of return difference calculations using SL Dep. The incremental tax return is not relevant to the decision. Extremely high interest rates would tend to favor the 10% ITC as described.

PEE Solutions Manual   Chapter 9

9-10
Income tax = $13,750 +0.34($85,000) +0.05($60,000)
         = $13,750 +$28,900 +$3,000 = $45,650
This is $45,650/$160,000 = 28.5% of the taxable income.
The applicable tax rate for economy studies should be the rate of the highest increment of income that is subject to tax. In this case, the rate is 39% for prospective investments yielding up to $175,000 per year taxable income. Beyond that level it is 34%.

9-11
The combined incremental rate is s +(1 - s)f
For $200,000, combined rate is 0.06 +(1.0 -0.06)(0.39) = 42.7%
For $400,000, combined rate is 0.06 +(1.0 -0.06)(0.34) = 38.0%
For a corporation that has a taxable income for federal purposes exceeding $335,000, the incremental federal rate will be 34%. In this case the taxable income for federal purposes is $376,000.

9-12
If the individual in this problem is single, the incremental federal tax rate is constant, 28%, for taxable income between $17,850 and $43,150. Thus the applicable federal tax rate will be 28%.
If the individual is married and files a joint return with his or her spouse, taxable income below $29,750 is taxed at 15% and income between $29,750 and $71,900 is taxed at 28%. If there is equal likelihood of the taxable income being below or above $29,750, the average of the two rates, 21.5%, would be a good estimate. Since the taxable income "generally falls between $22,000 and $40,000", it seems a little more likely that the actual rate would be 28%.

9-13
(a) 0 = -$50,000 +($16,000 -$3,000)(P/A,i%,5)
        +$15,000(P/F,i%,5)   i = 15.8%
(b) 0 = -$50,000 +$16,000(P/A,i%,5)
        +$15,000(P/F,i%,5)   i = 23.3%

9-14
Estimated extra annual income tax = 0.40($16,000 -$3,000 -$7,000)
                                  = $2,400
To compute prospective after-tax rate of return:
    0 = -$50,000 +($16,000 -$3,000 -$2,400)(P/A,i%,5)
        +$15,000(P/F,i%,5)   i = 9.7%

9-15

(a) (Headings same as Problem 9-1)

```
0    -$50,000                                              -$50,000
1    + 13,000   -$11,667   +$1,333    -$   533    + 12,467
2    + 13,000   -  9,333   + 3,667    -  1,467    + 11,533
3    + 13,000   -  7,000   + 6,000    -  2,400    + 10,600
4    + 13,000   -  4,667   + 8,333    -  3,333    +  9,667
5    + 13,000   -  2,333   +10,667    -  4,267    +  8,733
5    + 15,000                                              + 15,000
```

Rate of return after income taxes:
$0 = -\$47,667 + \$12,467(P/A,i\%,5) - \$933(P/G,i\%,5) + \$15,000(P/F,i\%,5)$
$i = 10.1\%$

The use of years-digits depreciation would increase the prospective rate of return from 9.7% to 10.1%

(b) Using the 1988 depreciation schedule for 5-year class assets, the cash flow table is:
(Headings same as Problem 9-1)

```
0   ($50,000)                                    ($50,000)
1   $13,000   ($14,000)   ($1,000)   $ 400       $13,400
2   $13,000   ($ 8,400)   ($1,000)   $ 400       $13,400
3   $13,000   ($ 5,040)   $4,600     ($1,840)    $11,160
4   $13,000   ($ 3,780)   $7,960     ($3,184)    $ 9,816
5   $13,000   ($ 3,780)   $9,220     ($3,688)    $24,312
5   $15,000
```

IRR A/T = 12.5%

9-16
Annual cost of rental = $\underline{\$25,000}$
Annual cost of ownership = $3,000 + $110,000(0.12) + $8,800
= $\underline{\$25,000}$

9-17
Annual cost with proposal rejected = $\underline{\$32,000}$
Annual cost with proposal accepted = $5,700 + $110,000(A/P,12%,10) + $6,120 = $\underline{\$31,288}$

PEE Solutions Manual   Chapter 9

9-18
Annual cost with proposal rejected = $36,300
Annual cost with proposal accepted = $10,000 +$110,000(A/P,12%,10)
   -[$2,064(P/F,12%,1) -$1,544(P/F,12%,2) -$4,096(P/F,12%,3)
   -$5,944(P/F,12%,4) -$6,692{(P/F,12%,5) +(P/F,12%,6)}
   -$6,736(P/F,12%,7) -$10,520{(P/F,12%,8) +(P/F,12%,9)
   +(P/F,12%,10)}](A/P,12%,10) = $34,378

9-19
A special difficulty is caused because the tax consequences extend beyond the 10-year study period that seems appropriate for a comparison of annual costs. The positive cash flow from tax savings of $1,760 a year for years 11-25 might be converted to an equivalent uniform annual figure for years 1-10. A possible method for this comparison is to calculate present worth at zero date as $1,760[(P/A,12%,25) -(P/A,12%,10)] = $3,860. This converts to a uniform figure for years 1-10 of $3,860(A/P,12%,10) = $683.
Annual cost for years 1-10 with proposal rejected = $36,300
Equivalent uniform annual cost for years 1-10 with proposal accepted
= $10,000 +$110,000(A/P,12%,10) +0.4($36,300 -$14,400) -$683 = $37,545

9-20
Annual cost with proposal rejected = $32,000
Annual cost with proposal accepted = $5,700 +$110,000(A/P,12%,10)
   +$1,720 +($10,520 -$1,720)[(P/A,12%,10)
   -(P/A,12%,5)](A/P,12%,10) = $5,700 + $110,000(0.17698) + $1,720
   +$8,800[5.650 - 3.605](0.17698) = $30,073

9-21
Annual cost with proposal rejected = $26,300
Annual cost with proposal accepted = ($110,000 -$44,000)(A/P,12%,10)
   +$10,520 = $66,000(0.17698) +$10,520 = $22,201

9-22
PW of rental = $25,000(P/A,11%,10) = $25,000(5.889) = $147,225
PW of ownership = $110,000 -$110,000(P/F,11%,10)
   +$3,000(P/A,11%,10) +$8,800(P/A,11%,10)
   = $110,000 -$110,000(0.3522) +$11,800(5.889) = $140,748

9-23
PW with proposal rejected = $32,000(P/A,11%,10)
   = $32,000(5.889) = $188,448
PW with proposal accepted = $110,000 +$5,700(5.889)
   +$6,120(5.889) = $179,608

PEE Solutions Manual   Chapter 9

9-24
   (See solution to 9-19)
   PW with proposal rejected = $32,000(P/A,11%,10)
      = $32,000(5.889) = $188,448
   PW with proposal accepted = $110,000 +$5,700(P/A,11%,10)
      +$8,760(P/A,11%,10) -$1,760[(P/A,11%,25)
      -(P/A,11%,10)] = $190,697

9-25
   PW with proposal rejected = $36,300(P/A,11%,10)
      = $36,300(5.889) = $213,771
   PW with proposal accepted = $110,000 +$10,000(P/A,11%,10)
      -$2,064(P/F,11%,1) +$1,544(P/F,11%,2) +$4,096(P/F,11%,3)
      +$5,944(P/F,11%,4) +$6,692[(P/F,11%,5) +(P/F,11%,6)]
      +$6,736(P/F,11%,7) +$10,520[(P/F,11%,8)
      +(P/F,11%,9) +(P/F,11%,10)] = $198,370

9-26
   PW with proposal rejected = $32,000(P/A,11%,10) = $188,450
   PW with proposal accepted = $110,000 +$5,700(P/A,11%,10)
      +$1,720(P/A,11%,10) +($10,520 -$1,720)[(P/A,11%,10)
      -(P/A,11%,5)] = $172,995

9-27
   PW with proposal rejected = $26,300(P/A,11%,10)
      = $26,300(5.889) = $154,880
   PW with proposal accepted = $110,000 - $44,000
      +$10,520(5.889) = $127,953

PEE Solutions Manual   Chapter 9

9-28(a)
(Headings same as Problem 9-1)

0	($ 110,000)			$ 11,000	($ 99,000)
1	$ 26,300	($ 11,000)	$ 15,300	($ 6,120)	$ 20,180
to	⋮	⋮	⋮	⋮	⋮
10	$ 26,300	($ 11,000)	$ 15,300	($ 6,120)	$ 20,180
Sums	($ 57,400)	($ 22,000)	$ 30,600	($ 1,240)	($ 58,640)
				IRR A/T =	15.6%

(b)
(Headings same as Problem 9-1)

0	($ 110,000)				($ 110,000)
1	$ 26,300	($ 44,000)	($ 17,700)	$ 7,080	$ 33,380
2	$ 26,300	($ 26,400)	($ 100)	$ 40	$ 26,340
3	$ 26,300	($ 15,840)	$ 10,460	($ 4,184)	$ 22,116
4	$ 26,300	($ 11,880)	$ 14,420	($ 5,768)	$ 20,532
5	$ 26,300	($ 11,880)	$ 14,420	($ 5,768)	$ 20,532
6	$ 26,300		$ 26,300	($ 10,520)	$ 15,780
7	$ 26,300		$ 26,300	($ 10,520)	$ 15,780
8	$ 26,300		$ 26,300	($ 10,520)	$ 15,780
9	$ 26,300		$ 26,300	($ 10,520)	$ 15,780
10	$ 26,300		$ 26,300	($ 10,520)	$ 15,780
SUMS	$ 153,000	($ 110,000)	$ 153,000	($ 61,200)	$ 91,800
				IRR A/T =	15.5%

PEE Solutions Manual   Chapter 9

9-29(a) and (b)
(Headings same as Problem 9-1)

0	($63,000)				($63,000)
1	$16,300	($4,200)	$12,100	($4,840)	$11,460
to					
15	$16,300	($4,200)	$12,100	($4,840)	$11,460
SUMS	$181,500	($63,000)	$181,500	($72,600)	$118,900
	25.0%	= IRR B/T		IRR A/T = 16.3%	

(c)
(Headings same as Problem 9-1)

0	($63,000)				($63,000)
1	$16,300	($8,398)	$7,902	($3,161)	$13,139
2	$16,300	($7,278)	$9,022	($3,609)	$12,691
3	$16,300	($6,308)	$9,992	($3,997)	$12,303
4	$16,300	($5,467)	$10,833	($4,333)	$11,967
5	$16,300	($4,739)	$11,561	($4,624)	$11,676
6	$16,300	($4,107)	$12,193	($4,877)	$11,423
7	$16,300	($3,560)	$12,740	($5,096)	$11,204
8	$16,300	($3,085)	$13,215	($5,286)	$11,014
9	$16,300	($2,677)	$13,623	($5,449)	$10,851
10	$16,300	($2,317)	$13,983	($5,593)	$10,707
11	$16,300	($2,008)	$14,292	($5,717)	$10,583
12	$16,300	($1,740)	$14,560	($5,824)	$10,476
13	$16,300	($1,508)	$14,792	($5,917)	$10,383
14	$16,300	($1,307)	$14,993	($5,997)	$10,303
15	$16,300	($1,133)	$15,167	($6,067)	$10,686
*	Loss on disposal =		($1,133)	$453	
SUMS	$181,500	($55,632)	$188,868	($75,547)	$106,406
	25.0%	= IRR B/T		IRR A/T = 17.0%	

*The indicated tax deduction for loss on disposal may or may not be permitted.

PEE Solutions Manual   Chapter 9

9-30(a) and (b)
(Headings same as Problem 9-1)

0	($63,000)				($63,000)
1	$16,300	($4,200)	$12,100	($4,114)	$12,186
to	⋮	⋮	⋮	⋮	⋮
15	$16,300	($4,200)	$12,100	($4,114)	$12,186
SUMS	$181,500	($63,000)	$181,500	($61,710)	$119,790
	25.0%	= IRR B/T		IRR A/T = 17.7%	

(c)
(Headings same as Problem 9-1)

0	($63,000)				($63,000)
1	$16,300	($8,398)	$7,902	($2,687)	$13,613
2	$16,300	($7,278)	$9,022	($3,067)	$13,233
3	$16,300	($6,308)	$9,992	($3,397)	$12,903
4	$16,300	($5,467)	$10,833	($3,683)	$12,617
5	$16,300	($4,739)	$11,561	($3,931)	$12,369
6	$16,300	($4,107)	$12,193	($4,146)	$12,154
7	$16,300	($3,560)	$12,740	($4,332)	$11,968
8	$16,300	($3,085)	$13,215	($4,493)	$11,807
9	$16,300	($2,677)	$13,623	($4,632)	$11,668
10	$16,300	($2,317)	$13,983	($4,754)	$11,546
11	$16,300	($2,008)	$14,292	($4,859)	$11,441
12	$16,300	($1,740)	$14,560	($4,950)	$11,350
13	$16,300	($1,508)	$14,792	($5,029)	$11,271
14	$16,300	($1,307)	$14,993	($5,098)	$11,202
15	$16,300	($1,133)	$15,167	($5,157)	$11,528
* Loss on disposal =			($1,133)	$385	
SUMS	$181,500	($55,632)	$188,868	($64,215)	$117,670
	25.0%	= IRR B/T		IRR A/T = 18.3%	

*The indicated tax deduction for loss on disposal may or may not be permitted.

PEE Solutions Manual   Chapter 9

9-31(a) and (b)
(Headings same as Problem 9-1)

0	($63,000)				($63,000)
1	$16,300	($4,200)	$12,100	($3,388)	$12,912
to	:	:	:	:	:
15	$16,300	($4,200)	$12,100	($3,388)	$12,912
SUMS	$181,500	($63,000)	$181,500	($50,820)	$130,680
	25.0%	= IRR B/T		IRR A/T = 19.0%	

(c)
(Headings same as Problem 9-1)

0	($63,000)				($63,000)
1	$16,300	($8,398)	$7,902	($2,213)	$14,087
2	$16,300	($7,278)	$9,022	($2,526)	$13,774
3	$16,300	($6,308)	$9,992	($2,798)	$13,502
4	$16,300	($5,467)	$10,833	($3,033)	$13,267
5	$16,300	($4,739)	$11,561	($3,237)	$13,063
6	$16,300	($4,107)	$12,193	($3,414)	$12,886
7	$16,300	($3,560)	$12,740	($3,567)	$12,733
8	$16,300	($3,085)	$13,215	($3,700)	$12,600
9	$16,300	($2,677)	$13,623	($3,814)	$12,486
10	$16,300	($2,317)	$13,983	($3,915)	$12,385
11	$16,300	($2,008)	$14,292	($4,002)	$12,298
12	$16,300	($1,740)	$14,560	($4,077)	$12,223
13	$16,300	($1,508)	$14,792	($4,142)	$12,158
14	$16,300	($1,307)	$14,993	($4,198)	$12,102
15	$16,300	($1,133)	$15,167	($4,247)	$12,370
	*Loss on disposal =		($1,133)	$317	
SUMS	$181,500	($55,632)	$188,868	($52,883)	$128,934
	25.0%	= IRR B/T		IRR A/T = 19.6%	

*The indicated tax deduction for loss on disposal may or may not be permitted.

PEE Solutions Manual   Chapter 9

9-32
(Example 9-1) Tax rate = 30%
    Annual taxes = $22,000(0.30) = $6,600
    i = ($22,000 -$6,600)/$110,000 = <u>14.0%</u>
(Example 9-2) Annual taxes = ($26,300 -$11,000)(0.30) = $4,590
Find i so that
    NPW = 0 = -$110,000 +($26,300 -$4,590)(P/A,i%,10)
    i = <u>14.8%</u>
(Example 9-3) Annual taxes: years 1 through 10 =
    ($26,300 -$4,400)(0.30) = $6,570
Years 11 through 25: ($0 -$4,400)(0.30) = -$1,320
    NPW = 0 = -$110,000 +$19,730(P/A,i%,10)
    +$1,320(P/A,i%,15)(P/F,i%,10)   i = <u>12.9%</u>
(Example 9-4) Obtain year-by-year taxable income from Figure 9-1.
Multiply each by 0.3 to obtain income tax for that year. The final
equation is:
    NPW = 0 = -$110,000 + $27,848(P/F,i%,1) +$25,142(P/F,i%,2)
    +$23,228(P/F,i%,3) +$21;842(P/F,i%,4) +$21,281(P/F,i%,5)
    +$21,281(P/F,i%,6) +$21,248(P/F,i%,7)
    +$18,410(P/A,i%,3)(P/F,i%,7)   i = <u>16.2%</u>
(Example 9-5) Annual tax, first 5 years = $4,300(0.3) = $1,290
    Annual taxes second 5 years = $26,300(0.3) = $7,890
    NPW = 0 = -$110,000 +$25,010(P/A,i%,5) +$18,410(P/A,i%,5)(P/F,i%,5)
    i = <u>16.2%</u>
(Example 9-6) Tax saving in year 0 = $110,000(0.3) = $33,000
    Annual taxes years 1-10 = $26,300(0.3) = $7,890
    NPW = -($110,000 -$33,000) +$18,410(P/A,i%,10)
    i = <u>20.1%</u>

9-33
The methodology in 9-33 is the same as 9-32, except that a tax rate
of 15% is substituted for 30% in all the equations. Only the equations
are shown here:
(Example 9-1) i = $18,700/$110,000 = 0.170
    i = <u>17.0%</u>
(Example 9-2) NPW = 0 = -$110,000 +$24,005(P/A,i%,10)
    i = <u>17.5%</u>
(Example 9-3) NPW = 0 = -$110,000 +$23,015(P/A,i%,10)
    +$600(P/A,i%,15)(P/F,i%,10)
    i = <u>18.2%</u>
(Example 9-4) NPW = 0 = -$110,000 +$27,074(P/F,i%,1)
    +$25,721(P/F,i%,2) +$24,764(P/F,i%,3) +$24,071(P/F,i%,4)
    +$23,791(P/F,i%,5) +$23,791(P/F,i%,6) +$23,774(P/F,i%,7)
    +$22,355(P/A,i%,3)(P/F,i%,7)
    i = <u>18.2%</u>
(Example 9-5) NPW = 0 = -$110,000 +$25,655(P/A,i%,5)
    +$22,355(P/A,i%,5)(P/F,i%,5)
    i = <u>18.2%</u>
(Example 9-6) NPW = 0 = -($110,000 -$16,500) +$22,355(P/A,i%,10)
    i = <u>20.1%</u>

9-34
After-tax annual return on the telephone bond
= $550 -$550(0.28) = $396
Income from tax exempt municipal bond = $375
The telephone bond is more attractive by $21 a year. Comparison could have been by comparing the after-tax return on the telephone bond with that of the tax exempt bond:
11.0%(1-0.28) = 7.92% which is > 7.5%

9-35
Gross income as a tax base disregards the expenses necessary to produce income. Therefore it cannot reflect capacity to make tax payments. Moreover, a tax on gross income is socially undesirable because it gives an unfair competitive advantage to those integrated manufacturing enterprises that start with producing raw materials and end with marketing a finished product.

9-36
Public utility bonds:
NPW = 0 = -$15,000 +[$1,980(1 -0.35)(P/A,i%,8)] +$15,000(P/F,i%,8)
i = **8.58%**
Municipal bonds:
NPW = 0 = -$15,000 +$600(P/A,i%,8) +[$20,000 -$5,000(0.28)](P/F,i%,8)
i = **6.39%**

9-37
David's combined applicable tax rate = 0.11 +(0.89)(0.33) = **40.4%**
Net cost of the charitable gift = $2,000 -$2,000(0.404) = **$1,192**

9-38

(a) Rate of return before income taxes:
NPW = 0 = -$20,000 +$5,600(P/A,i%,10)   i = <u>25.0%</u>

(b) After-tax cash flows: (Headings same as Problem 9-1)

```
 0     -$20,000                                    +$2,000    -$18,000
 1     + 5,600    -$5,600       $    0                  0     + 5,600
2-10   + 5,600    - 1,600      +$4,000             - 1,800    + 3,800
```

0 = -$18,000 +$1,800(P/F,i%,1) +$3,800(P/A,i%,10)   i = <u>19.0%</u>

(c)

```
 0     -$20,000                                    +$2,000    -$18,000
1-10   + 5,600    -$2,000      +$3,600             - 1,620    + 3,980
```

0 = -$18,000 +$3,980(P/A,i%,10)   i = <u>17.8%</u>

(d) Differences in cash flow are +$1,620 at year 1 and -$180 a year from years 2-10. With an i% of 10%, the present worth of this at date 0 is $1,620(0.909) -$180(6.144 -0.909) = <u>$530</u>.

9-39

(Headings same as Problem 9-1)

0	($ 20,000)			($ 20,000)	
1	$ 5,600	($ 5,720)	($ 120)	$ 48	$ 5,648
2	$ 5,600	($ 4,080)	$ 1,520	($ 608)	$ 4,992
3	$ 5,600	($ 2,920)	$ 2,680	($ 1,072)	$ 4,528
4	$ 5,600	($ 2,080)	$ 3,520	($ 1,408)	$ 4,192
5	$ 5,600	($ 1,740)	$ 3,860	($ 1,544)	$ 4,056
6	$ 5,600	($ 1,740)	$ 3,860	($ 1,544)	$ 4,056
7	$ 5,600	($ 1,720)	$ 3,880	($ 1,552)	$ 4,048
8	$ 5,600		$ 5,600	($ 2,240)	$ 3,360
9	$ 5,600		$ 5,600	($ 2,240)	$ 3,360
10	$ 5,600		$ 5,600	($ 2,240)	$ 3,360
Sums	$ 36,000	($ 20,000)	$ 36,000	($ 14,400)	$ 21,600
	25% = IRR B/T			IRR A/T = 18.2%	

In Problem 9-38, the after-tax rate of return with the investment tax credit and additional first-year depreciation was 19.0%. Under the 1986 tax laws with the 7-year class depreciation, the rate of return declines to 18.2%. Thus, in this case, the investor was not quite as well off under the new tax laws.

PEE Solutions Manual   Chapter 9

9-40
Combined applicable tax rate, ordinary income = 0.08 +(0.92)(0.28)
          = <u>33.8%</u>
Capital gains tax rate = 0.06 +(0.94)(0.28) = <u>32.3%</u>
Company bonds: NPW = 0 = -$5,000 +($275 -$92.95)(P/A,i%,10)
   +$5,000(P/F,i%,10)
   i = <u>3.641%</u> for half year.   Effective annual rate = <u>7.41%</u>

Municipal bonds: NPW = 0 = -$5,000 +$180(P/A,i%,10)
   +[$6,000 -$1,000(0.323)](P/F,i%,10)
   i = <u>4.692%</u> for half year.   Effective annual rate = <u>9.60%</u>

9-41
EUAC at 8% if renting = -$11,400 +$10,000(0.08)(1.0 -0.33)
     = -$11,400 +$536 = -$10,854
Ownership: The table below shows that, since interest on loan and property taxes are deductible in calculating Ron's taxable income, the net annual expenditure is around $8,275 each year. Then he anticipates a capital gain of $7,000 taxed at 28% and return of his down payment of $10,000. That leads to an approximate EUAC of only $6,672. That provides a lot of incentive to buy rather than rent. <u>Note</u> that annual end of year mortgage and rent payments were assumed rather than actual monthly payments.

Year	Own'ship	Annual Disburse't	Interest & Prop. Tax	Tax Saving	Net Cash Flow
0	($10,000)				($10,000)
1	($8,906)	($2,400)	($9,250)	$3,053	($8,254)
2	($8,906)	($2,400)	($9,178)	$3,029	($8,277)
3	($8,906)	($2,400)	($9,098)	$3,002	($8,304)
3	$17,360				$17,360
				NPV @ 8%	($17,195)
				EUAC @ 8%	($6,672)

$2,191 will have been paid off on the $75,000 mortgage in three years. He will have $17,231 cash after sale of the property in 3 years.

9-42

(a) NPW = 0 = -$2,500 +$28.15(0.65)(P/A,i%,32)
   +[$3,400 -$252](P/F,i%,32)
   i = **1.39%** for quarter year. Effective annual rate = **5.68%**

(b) NPW = 0 = -$2,500 +$28.15(0.65)(P/A,i%,28)
   +[$3,200 -$98](P/F,i%,28)
   i = **1.44%** for quarter year. Effective annual rate = **5.89%**

Mary would have earned a slightly higher return by selling earlier. If invested at 6%, compounded quarterly and taxed at 35%, the quarterly after-tax interest rate = 1.5(0.65) = 0.975%; she would have $3,102(F/P,0.975%,4) = $3,224.76 on December 29, 1987, as opposed to $3,148.00 from holding the stock the extra year.

PEE Solutions Manual

CHAPTER 10

Increment Costs, Economic Sizing, Sunk Costs, and Interdependent Decisions

General Notes:

This chapter explains several important principles that are frequently violated by otherwise knowledgeable businessmen. They deserve considerable emphasis in the classroom.

The tendency to use averages is very strong; yet, as the very simple examples of this chapter show, it can lead to very bad decisions. The concept of incremental costs is obvious when a little thought is given to it. However, accounting practices frequently lead engineers and businessmen to accept and use cost accounting data that include allocated costs that are irrelevant to the decision at hand. The question, "What are the actual differences that will occur with or without the proposed action?" will help to show the importance of understanding incremental cost analysis.

The question just suggested can also help students understand the sunk cost concept. If they are forced to determine only those differences that lie in the future, they will quickly grasp the idea that a decision now cannot change things that occurred in the past, that a decision now only affects the future.

Most of us have a psychological block against admitting even the appearance that we have made a mistake. Frequently we try to cover up such apparent mistakes by introducing some of the costs of those mistakes into the charges against future alternatives. This kind of error results from not understanding the sunk cost concept and the true situation one encounters when a decision is to be made.

10-1
Only the final 20 kw-hr of the 500 are priced at 8.1¢. The remaining 80 kw-hr of the proposed 100 kw-hr reduction are priced at 5.9¢. The reduction in the bill will be 20($0.081) + 80($0.059) = $1.62 + $4.72 = $6.34. When $6.34 is divided by 100 hw-hr the quotient is 6.34¢/kw-hr. Nevertheless, this average cost per kw-hr saved is not a unit cost that is significant for the purposes of any decision; the 8.1¢ and 5.9¢ figures are the relevant unit costs for decision making.

10-2
This calculated unit cost of 23.3¢ per kw-hr has no relevance to any decision. The relevant costs are $1.82 per kw of demand on the system and 5.1¢ per kw-hr for energy consumed.

10-3

(a) Demand charge:
    First 5,000 kw $11,700
    Next 4,500 kw @ $2.60   <u>11,700</u>
                                                           $23,400

Energy charge:
    1,800,000 kw-hr @ 4.9¢* $88,200

Monthly bill   <u>$111,600</u>

Average cost is $111,600 ÷ 1,800,000 = 6.2¢/kw-hr

*The 4.9¢ price applies to all kw-hr up to 500,000 + 300(9,500) = 2,850,000 kw-hr. Because the actual consumption is less than this, all units purchased are priced at 4.9¢.

(b) Extra demand charge will be:
    500 kw @ $2.60 $1,300
    500 kw @ $1.90   <u>950</u>
                                                               $2,250

Extra energy charge will be:
    160,000 kw-hr @ 4.9¢ $7,840

Total added cost   <u>$10,090</u>

The average cost of the extra energy will be $10,090 ÷ 160,000 = 6.31¢/kw-hr

10-4

Demand charge still will be $23,400

Energy charge now is
    1,170,000 kw-hr @ 4.9¢   <u>57,330</u>
Total monthly bill $80,730

Average cost is $80,730 ÷ 1,170,000 = 6.9¢/kw-hr

PEE Solutions Manual   Chapter 10

10-5
It was calculated in the solution to Problem 10-3 that the energy charge per kw-hr would decline from 4.9¢ to 4.2¢ for all kw-hr purchased in excess of 2,850,000 per month. Because the maximum demand will be unchanged by the change to two-shift operation, this figure is still appropriate. The extra shift increases monthly energy consumption from 1,800,000 to 3,200,000 kw-hr. Of this extra 1,400,000 kw-hr, 1,050,000 will be purchased at 4.9¢ and the remaining 350,000 at 4.2¢.

1,050,000 @ 4.9¢	$51,450
350,000 @ 4.2¢	14,200
Total	$66,150

Because the demand charge will be unchanged, this $66,150 will be the increase in the monthly electric bill. The average cost of the energy added by the second shift will be $66,150 ÷ 1,400,000 = 4.73¢/kw-hr.

10-6
$a = \$2.60(0.00302)(0.20182) = \$0.001584$

$b = \dfrac{(40)^2(3,500)(\$0.106)(10,580)}{1,000} = \$6,280,288$

$x_e = \sqrt{\dfrac{b}{a}} = \sqrt{\dfrac{\$4,265,856}{\$0.001158}} = 62,967$ circular mils

This falls between size 2 (66,400 circular mils) and size 3 (52,600 circular mils). Size 2 should be chosen.

10-7

A	Size of wire (AWG)	00	0	1	2	3
B	Weight of wire in lb.	403	319	253	201	159
C	Investment in wire	$1,047.80	$829.40	$657.80	$522.60	$413.40
D	Resistance in ohms	0.0795	0.100	0.126	0.159	0.201
E	Power loss - kw	0.1272	0.1600	0.2016	0.2544	0.3216
F	Annual energy loss in kw-hr	445	560	706	890	1,126
G	Annual investment charges @ 20.182%	$211.47	$167.39	$132.76	$105.47	$83.43
H	Annual cost of lost energy @ 10.6¢/kw-hr	$47.17	$ 59.36	$74.84	$ 94.34	$119.36
I	Equivalent uniform annual cost assumed to be variable with wire size	$258.64	$226.75	$207.60	$199.81	$202.79

Size 2 has the lowest cost.

PEE Solutions Manual   Chapter 10

10-8

	A Size of wire	00	0	1	2	3
	B CR @ 7% (0.08581)	$60.52	$47.90	$37.99	$30.18	$23.88
	C Property tax (0.0175)	12.34	9.77	7.75	6.16	4.87
	D Depreciation	28.21	22.33	17.71	14.07	11.13
	E Cost of lost energy	45.93	57.75	72.77	91.85	116.11
	F Deductions from taxable income (C + D + E)	86.48	89.85	98.23	112.08	132.11
	G Extra annual income taxes above size 3	18.25	16.90	13.55	8.01	0.00
	H Equivalent uniform annual cost assumed to be variable with wire size	$137.04	$132.32	$132.06	$132.20	$144.86

Size 1 now shows the lowest annual cost. However, the differences in annual costs among the different wire sizes are considerably less than in the before-tax analysis of Table 10-2.

10-9

	A Size of wire	00	0	1	2	3
	B Weight, lbs.	403	319	253	201	159
	C Investment @$2.60/lb	$1,047.80	$829.40	$657.80	$522.60	$413.40
	D Resistance, ohms	0.0795	0.100	0.126	0.159	0.201
	E Power loss, 60A, in kw,	0.2862	0.3600	0.4536	0.5724	0.7236
	F Annual energy loss, kw. hr.	1,717	2,160	2,722	3,434	4,342
	G Investment charges @ 18.5%	$193.84	$153.44	$121.69	$96.68	$76.48
	H Cost of lost energy	120.19	151.20	190.54	240.38	303.94
	I Total annual cost assumed to be variable with the size of wire	$314.03	$304.64	$312.23	$337.06	$380.42

Size 0 is most economical at the designated load.

10-10

You have already spent $140. Nothing you can do now will change that. It is past and is a sunk cost. If you can assure a sale, and a commission income of $200, by spending some additional money, you will improve your situation if you can assure the sale by spending any amount less than $200. Thus, at this point in your decision process, the past expenditure of $140 is completely irrelevant except as it helps you predict the probability of success if additional money is spent.

PEE Solutions Manual   Chapter 10

10-11
   If there were no income tax considerations, Frank and James ought to reason alike; a major element in the decision ought to be a forecast of the future price of XY stock. But it is possible that Frank might be inclined to borrow the $20,000 because of being reluctant to take a loss on his stock, and that James might be inclined to sell his stock because of being pleased at making a profit.
   A sophisticated look at income tax considerations ought to make Frank more willing to sell his stock and make James less willing. Frank would have a deductible capital loss that would reduce his current payment of income taxes. In contrast, James would have a taxable capital gain that would increase his current tax payments.

10-12
   After each such sale and replacement of a unit the merchant had the same stock and $5 less cash. It is evident that he was $5 worse off after each such transaction regardless of what his accounts might have indicated.

10-13
   Both friends are wrong. The alternatives are:

	Go to jail	Do not go to jail
Pay fine	--	-$20
Write story	+$25	--
	+$25	-$20

The difference between these two alternatives is $45.
   (Note: Whose viewpoint and when the decision was made will be raised by some students. Did Steve know when he decided to go to jail rather than pay his fine that he could write and sell an article on the experience? The total difference of $45 is an after-the-fact analysis, and is the correct answer to the question raised in the problem, but it is not necessarily the basis for Steve's decision when he elected to go to jail.)

PEE Solutions Manual   Chapter 10

10-14
Demand charge:
    First 40 kw $91.00
    Next 240 kw @ $1.99   477.60
                                          $568.60

Energy charge:
    55,000 kw-hr @ 5.1¢ $2,805.00
    Monthly bill $3,373.60

Average cost = $3,373.60 ÷ 55,000 = 6.13¢/kw-hr

The proposed new machinery will increase the monthly bill as follows:

Demand charge:
    20 kw @ $1.99 $39.80
    30 kw @ $1.82   54.60
                                            $94.40

Energy charge:
    2,000 kw-hr @ 5.1¢ $102.00

Addition to monthly bill $196.40

Average cost of energy added is $196.40 ÷ 2,000 = 9.82¢/kw-hr

10-15
Reduction in demand charge = 40 @ $1.99   = $79.60
Reduction in energy charge = 1,500 @ 5.1¢ = 76.50
    Total reduction in bill $156.10
This is an average saving of 10.41¢/kw-hr

10-16

	Georgia (2,700 T)	Florida (3,900 T)
Total Cost: $135(2,700) $142(3,900)	$364,500	$553,800
Less Fixed Cost:	96,930	196,170
Total Variable Cost:	$267,570	$357,630
Waste Paper Cost: $34(1,440) +$43.00(0.8)(900) $36.50(0.8)(3,900)	79,920	113,880
Total Variable Cost other than W.P.	$187,650	$243,750
Per Ton	$ 69.50	$ 62.50

In computing the waste paper cost for the Georgia plant, the 1,440 tons available at $34 would produce 1,440/0.8 = 1,800 tons of finished paperboard. The remaining 900 tons of finished paperboard would be made from 900(0.8) = 720 tons of waste paper at $43.00 per ton.

10-16 (cont)
Production scheduling in the short run will be done without regard to fixed costs:
Variable costs:
Georgia plant
$69.50 +$34(0.8) = $96.70/T up to 1,800 T/mo output
$69.50 +$43(0.8) = $103.90/T above 1,800 T/mo output
Florida plant
$62.50 +$36.50(0.8) = $91.70/T up to 4,000 T/mo output
$62.50 +$43(0.8) = $96.90/T above 4,000 T/mo output

The order of scheduling priority should be:
Florida plant up to 4,000 tons output
Georgia plant up to 1,800 tons output
Florida plant up to 6,000 tons output (capacity) in that order
Georgia plant up to 3,600 tons output (capacity)

(a) For a total production of 6,600 tons, 1,800 tons should be scheduled for the Georgia plant and 4,800 tons for the Florida plant. This will result in a cost saving of $6,820 per month over present operating costs.

(b) For a total production of 9,000 tons:
```
Georgia  -  3,000              $395,670
Florida  -  6,000               756,770
Total cost/mo.               $1,152,440
```

10-17

(a)	San Francisquito	Los Trancos
Present Output	1,600 T	2,000 T
Avg. Cost/T	$256	$248
Total Cost	$409,600	$496,000
Fixed Cost	73,600	44,000
Variable Cost	336,000	452,000
Var. Cost/T	$210	$226

$16.00/T can be saved by shifting production from the L.T. to the S.F. plant.

(b) The additional 400 T/mo will cost $84,000/mo if produced at S.F.; it will cost $90,400/mo if produced at L.T. The monthly saving by producing at S.F. would be $6,400.

10-18
Both viewpoints are wrong. Jones and Smith now have the equivalent of $800, or each unit is worth $0.10. A more reasonable viewpoint, therefore, is that Jones or Smith should not spend a currency unit unless they believe they are getting $0.10 worth of goods for it. The values of the currency units before the change in rates are not relevant.

10-19

If it is assumed that an annual cost comparison is to be made for 9 years, the following criticisms are relevant.
1. Depreciation (capital recovery) has been counted twice: once as a charge against continued ownership and again as a credit toward rental.
2. The charge for depreciation (capital recovery) against continued ownership was based on the original cost when it should have been based on net realizable salvage value.
3. The $120 unusual repair cost was treated as an annual cost when it should have been amortized over the remaining life (unless there is reason to suspect a recurrence of the uninsured damage.)

A correct cost analysis yields the following on a before-tax basis:
Cost of renting = $12(2)(12) = $288

Continued ownership, EUAC for 9 years
($720 +$120)(A/P,i%,9) +$45 +$40

@ 0% interest (which is equivalent to S.L. depreciation over remaining life) =
$840/9 +$85 = $178.33
@ 15% $840(0.20957) +$85 = $261.04
@ 25% $840(0.28876) +$85 = $327.56

The decision favors continued ownership at a before-tax RoR up to 19%; above that rate, renting is more economical.

10-20

Indirect manufacturing expense is made up of fixed costs plus indirect variable costs. Treating it as all variable cost gives an incorrect effect to fixed costs. Since production scheduling assignments should always be based on variable cost analysis, consideration of total accounting costs may lead to improper machine assignment at all levels of output. Manual and partly mechanized (general purpose) operations will have an improper advantage over highly mechanized or automated (special purpose) operations.

Hence, for production scheduling purposes, an attempt should be made to estimate "variable burden" (power, water, scrap, maintenance, housekeeping, inspection, etc.) and to include these costs with the direct variable costs of setup, direct labor, and materials.

PEE Solutions Manual																Chapter 10

10-20 (cont)
These comments apply to all levels of plant output up to plant capacity, including conditions (a) and (b). However, under condition (c), the plant working at or near capacity with a large volume of unfilled orders, some modification of the above comments is appropriate. In considering unfilled orders, the alternatives are:

I. Let the customers wait until plant capacity becomes available, and perhaps incur their wrath;

II. Process the backlog as soon as possible by:
  (a) Scheduled overtime at premium pay;
  (b) Purchasing additional equipment and hiring the needed labor;
or (c) Subcontracting out the work as possible.

In comparing the relative cost of these alternatives, the next avoidable increment of cost is the total cost of new equipment or subcontracting, rather than just the variable cost of owned equipment.

10-21
It is unfortunate that the management writer confused the rate of return methodology per se with an improper application. Were the analysis properly structured, the rate of return approach would be a useful guide to decision making.

Consider the following cash flow table of the alternatives available to John Doe, and the differences between these alternatives:

Year	Buy Concession	Do Not Buy Concession	Difference
0	-$900	+$100	-$1000
1-3	+ 450		+ 450
3	+ 900		+ 900

Roe's offer decreased the relative attractiveness of Doe's initial offer (as it was intended to do), but the net superiority of the initial concession offer still remains attractive, yielding 42.8% as determined from:

$$0 = -\$1{,}000 + \$450(P/A, i\%, 3) + \$900(P/F, i\%, 3)$$

It is not proper to consider the rate of return for mutually exclusive alternatives separately, since the receipt of $100 for not buying the concession is dependent on the existence of the other alternative, and must be compared with it. That is, $100 will be received <u>provided</u> one forgoes the attractive cash flow of the concession.

PEE Solutions Manual

CHAPTER 11

Economy Studies for Retirement and Replacement

11-1
Defender Analysis

year	Cash flow before income taxes	Write-off of initial outlay for Tax purposes	Influence on taxable income	Influence of income taxes on cash flow −0.40C	Cash flow after income taxes (A + D)
0	($ 7,000)				($ 7,000)
1	($ 12,500)	($ 2,500)	($ 15,000)	$ 6,000	($ 6,500)
2	($ 12,500)	($ 2,500)	($ 15,000)	$ 6,000	($ 6,500)
3	($ 12,500)	($ 2,500)	($ 15,000)	$ 6,000	($ 6,500)
4	($ 12,500)	($ 2,500)	($ 15,000)	$ 6,000	($ 6,500)
5	($ 12,500)	($ 2,500)	($ 15,000)	$ 6,000	($ 4,500)
5	$ 2,000				
				EUAC =	($ 8,292)

Challenger Analysis

year					
0	($ 36,000)				($ 36,000)
1	($ 2,500)	($ 10,286)	($ 12,786)	$ 5,114	$ 2,614
2	($ 2,500)	($ 7,347)	($ 9,847)	$ 3,939	$ 1,439
3	($ 2,500)	($ 5,248)	($ 7,748)	$ 3,099	$ 599
4	($ 2,500)	($ 3,748)	($ 6,248)	$ 2,499	($ 1)
5	($ 2,500)	($ 3,124)	($ 5,624)	$ 2,249	($ 251)
6	($ 2,500)	($ 3,125)	($ 5,625)	$ 2,250	($ 250)
7	($ 2,500)	($ 3,121)	($ 5,621)	$ 2,248	($ 252)
8	($ 2,500)		($ 2,500)	$ 1,000	($ 1,500)
9	($ 2,500)		($ 2,500)	$ 1,000	($ 1,500)
10	($ 2,500)		($ 2,500)	$ 1,000	($ 1,500)
				NPV @15%	($ 33,861)
				EUAC =	($ 6,747)

PEE Solutions Manual   Chapter 11

11-2
Defender (Headings same as Problem 11-1)

0	($ 1,500)		$ 510	($ 990)
1	($ 3,000)	($ 3,000)	$ 1,020	($ 1,980)
2	($ 3,000)	($ 3,000)	$ 1,020	($ 1,980)
3	($ 3,000)	($ 3,000)	$ 1,020	($ 1,980)
4	($ 3,000)	($ 3,000)	$ 1,020	($ 1,980)
5	($ 3,000)	($ 3,000)	$ 1,020	($ 1,980)
			NPV @12% =	-8,127
			EUAC =	($ 2,255)

Challenger (Headings same as Problem 11-1)

0	($ 22,000)			($ 22,000)
1		($ 8,800)	($ 8,800)	$ 2,992	$ 2,992
2		($ 5,280)	($ 5,280)	$ 1,795	$ 1,795
3		($ 3,168)	($ 3,168)	$ 1,077	$ 1,077
4		($ 2,376)	($ 2,376)	$ 808	$ 808
5		($ 2,376)	($ 2,376)	$ 808	$ 808
			NPV @12%	($ 16,159)
			EUAC =	($ 4,483)

11-3
Let X = the required selling price of old unit with 2-year life:
    -0.66X(A/P,12%,2) -$1,980 = -$4,483

X = ($4,483 -$1,980)/[0.5917(0.66)] = $6,409 to break even vs. the challenger

PEE Solutions Manual   Chapter 11

11-4
Defender (Headings same as Problem 11-1)

0	($ 3,000)		($ 9,000)	$ 3,060	$ 60
1	($ 19,500)	($ 1,200)	($ 20,700)	$ 8,280	($ 11,220)
2	($ 19,500)	($ 1,200)	($ 20,700)	$ 8,280	($ 11,220)
3	($ 19,500)	($ 1,200)	($ 20,700)	$ 8,280	($ 11,220)
4	($ 19,500)	($ 1,200)	($ 20,700)	$ 8,280	($ 11,220)
5	($ 19,500)	($ 1,200)	($ 20,700)	$ 8,280	($ 11,250)
5	$ 1,500		($ 4,500)	($ 1,530)	
	($ 50,898) = NPV @ 30%			NPV @ 15%	($ 37,686)
	($ 20,898) =EUAC			EUAC =	($ 11,242)

Challenger (Headings same as Problem 11-1)

0	($ 45,000)				($ 45,000)
1	($ 8,000)	($ 12,857)	($ 20,857)	$ 8,343	$ 343
2	($ 8,000)	($ 9,184)	($ 17,184)	$ 6,873	($ 1,127)
3	($ 8,000)	($ 6,560)	($ 14,560)	$ 5,824	($ 2,176)
4	($ 8,000)	($ 4,685)	($ 12,685)	$ 5,074	($ 2,926)
5	($ 8,000)	($ 3,905)	($ 11,905)	$ 4,762	$ 9,317
5	$ 15,000	($ 7,190)	$ 7,190	($ 2,445)	
	($ 60,445) = NPV@30%			NPV @15% =	-44,025
	($ 24,818) = EUAC			EUAC =	($ 13,133)

PEE Solutions Manual   Chapter 11

11-5
Old Flume: EUAC

Maintenance $6,800
Higher water loss:
(0.15 -0.03)(10M)($30.75/M)(365)   13,469
                                                                      $20,269

New Flume:

CR = $185,000(A/P,8%,20) = $185,000(0.10185) $18,842
Maintenance   2,000
                                                                      $20,842

The "Old Flume" analysis is a next-year cost; appropriate under the assumptions that the flume has no salvage value and will not leak in the future less than it will next year.
While it is most economical to keep the old flume for the present, the costs are so close that irreducibles will likely swing the decision in favor of replacement. Possible future increases in the price of electricity also play an important part in the decision.

11-6
Plan A:

CR = -$16,000(A/P,25%,10) =   -$4,481
Maintenance =   -   350
Energy = -(800)(41)($0.065) =   - 2,132
                                                               -$6,963

Plan B:

CR: = -($2,000 +$9,000)(A/P,25%,10) =   -$3,080
Maintenance = -($300 +$400) =   -   700
Energy = -2(1,200)(23)($0.065) =   - 3,588
                                                               -$7,368

Replacement should be made now.

111

PEE Solutions Manual   Chapter 11

11-7
Using the information in Problem 11-1, Defender, and calculating the book value at the end of 8 years and the tax savings if disposed of then ($4,000), the EUAC equation becomes:

$EUAC_{6th\ yr.}$ = -$7,000(A/P,15%,3) -$6,500 +$4,000(A/F,15%,3)

= -$3,066 -$6,500 +$1,152 = -$8,414

Replacement should be made now. As was to be expected, the cost of continuing the old equipment in service three more years was greater than the EUAC for continuing for 5 years because the initial loss of tax savings had to be absorbed in three years without a compensating increase in salvage value at end of three years.

11-8
P = $15,000, i = 20% before taxes
AC_3 = ($15,000 -$5,800)(A/P,20%,3) +[$276(P/F,20%,1)
        +$464(P/F,20%,2) +$1,230(P/F,20%,3)](A/P,20%,3) = $4,968

t	S_t	A_t	MC_t	$\Sigma PW(MC_t)$	AC_t
0	$15,000				
1	-	-276			
2	-	464			
3	5,800	1,230		$10,462	$4,968
4	4,400	1,980	$4,540	12,651	4,887
5	3,200	3,050	5,130	14,713	4,920
6	2,200	5,250	6,890	17,021	5,118

CAC = $4,887
Defender next-year cost = MC_5 +$300 = $5,130 +$300 = $5,330 = DAC
Replace now.

PEE Solutions Manual   Chapter 11

11-9
Plan A (Headings same as Problem 11-1)

0	($ 16,000)				($ 16,000)
1	($ 2,482)	($ 4,576)	($ 7,058)	$ 2,823	$ 341
2	($ 2,482)	($ 3,264)	($ 5,746)	$ 2,298	($ 184)
3	($ 2,482)	($ 2,336)	($ 4,818)	$ 1,927	($ 555)
4	($ 2,482)	($ 1,664)	($ 4,146)	$ 1,658	($ 824)
5	($ 2,482)	($ 1,392)	($ 3,874)	$ 1,550	($ 932)
6	($ 2,482)	($ 1,392)	($ 3,874)	$ 1,550	($ 932)
7	($ 2,482)	($ 1,376)	($ 3,858)	$ 1,543	($ 939)
8	($ 2,482)		($ 2,482)	$ 993	($ 1,489)
9	($ 2,482)		($ 2,482)	$ 993	($ 1,489)
10	($ 2,482)		($ 2,482)	$ 993	($ 1,489)
Sums	($ 40,820)	($ 16,000)	($ 40,820)	$ 16,328	($ 24,492)
				NPV =	($ 16,674)
				EUAC =	($3,322)

Plan B (Headings same as Problem 11-1)

0	($ 11,000)				($ 11,000)
1	($ 4,288)	($ 4,133)	($ 11,000)	$ 4,400	$ 112
2	($ 4,288)	($ 2,369)	($ 8,421)	$ 3,368	($ 920)
3	($ 4,288)	($ 1,847)	($ 6,657)	$ 2,663	($ 1,625)
4	($ 4,288)	($ 1,469)	($ 6,135)	$ 2,454	($ 1,834)
5	($ 4,288)	($ 1,316)	($ 5,757)	$ 2,303	($ 1,985)
6	($ 4,288)	($ 1,316)	($ 5,604)	$ 2,242	($ 2,046)
7	($ 4,288)	($ 1,307)	($ 5,604)	$ 2,242	($ 2,046)
8	($ 4,288)	($ 533)	($ 5,595)	$ 2,238	($ 2,050)
9	($ 4,288)	($ 533)	($ 4,821)	$ 1,928	($ 2,360)
10	($ 4,288)	($ 533)	($ 4,821)	$ 1,928	($ 2,360)
Sums	($ 53,880)	($ 15,356)	($ 64,415)	$ 25,766	($ 28,114)
				NPV @15% =	($ 15,896)
				EUAC =	($3,167)

Plan B is preferable.

11-10
From Table 11-2, using i* = 15%

t	(n)	MC_t	$\Sigma PW(MC_t)$	AC_t
1	2	$13,400	$11,653	$13,400
2	3	12,950	21,444	13,191
3	4	13,000	29,992	13,136
4	5	13,625	37,783	13,234

The defender advantage cost is $13,136, the average cost for 3 years more service. While this may imply a remaining economic life of 3 years, that would yield a total economic life of only 4 years. While the DAC is $13,136, this is for comparative purposes only against the CAC of $15,143. The comparison simply means do not replace now. Replacement is made when the DAC exceeds the CAC.

11-11

Four Year Life: Annual Cost

CR = $46,000(A/P,18%,4) -$20,000(A/F,18%,4)
   = $46,000(0.37174) -$20,000(0.19174) = $13,265
Disbursements = $6,000 +$500(A/G,18%,4)
   = $6,000 +$500(1.295) =   6,648
                                                                                                 $19,913

Eight Year Life:

CR = $46,000(A/P,18%,8) -$10,000(A/F,18%,8)
   = $46,000(0.24524) -$10,000(0.06524) = $10,629
Overhaul = $8,500(P/F,18%,4)(A/P,18%,8)
   = $8,500(0.5158)(0.24524) =   1,075
Disbursements = $6,000 +$500(A/G,18%,8)
   = $6,000 +$500(2.656) =   7,328
                                                                                                 $19,032

Twelve Year Life:

CR = $46,000(A/P,18%,12) -$5,000(A/F,18%,12)
   = $46,000(0.20863) -$5,000(0.02863) = $ 9,454
Overhaul = [$8,500(P/F,18%,4) +$10,000(P/F,18%,8)]
   (A/P,18%,12) = [$8,500(0.5158) +$10,000(0.2660)]
   (0.20863) =   1,470
Disbursements = [$6,000(P/A,18%,12) +$500(P/G,18%,8)
   +$7,000(P/A,18%,4)(P/F,18%,8)](A/P,18%,12)
   = [$6,000(4.793) +$500(10.829)
   + $7,000(2.690)(0.2660)](0.20863) =   8,174
                                                                                                $19,098

PEE Solutions Manual   Chapter 11

11-12

New tractor (the challenger) for eight year life:
The CAC = $19,032 (from Problem 11-11)

8-yr-old tractor (the defender):
CR = ($10,000 +$10,000)(A/P,18%,4)
  -$5,000(A/F,18%,4) = $ 6,476
Annual disbursements =   13,000
DAC = $19,476
Make the replacement now.

11-13

New tractor (the challenger) for eight year life:
CAC = $19,032 (from Problem 11-11)

4-year-old tractor (the defender):
CR = ($20,000 +$8,500)(A/P,18%,4) -$10,000(A/F,18%,4) = $8,677
Annual disbursements = $8,000 +$500(A/G,18%,4)
   = $8,000 +$500(1.295) =   8,648
DAC = $17,325
Retain the defender.
In order for replacement to be attractive, the present salvage must be at least:
   $P(A/P,18%,4) = $19,032 -$8,648 -$1,242
      from which $P = **$24,593**

11-14

(a) Annual disbursements have averaged: 1-$175; 2-$224; 3-$336; 4-$511; 5-$735; 6-$1,106. If this trend continues, it should not be replaced _now_ because the next-year cost will probably be about $1,512, which is less than the minimum EUAC for an average replacement (CAC = $2,048).

(b) Since the last-year disbursements ($2,120) exceed the CAC ($2,048), the machine should be replaced despite the fact it is only 5 years old.

(c) Since a general inflation increases all costs proportionately, the economic time of replacement will remain the same. All of the costs in the table will simply be increased by a factor of 1.3.

(d) If it is assumed that all four existing machines are 5 years old, presumably they would have one year of economic life remaining. However, since the four replacements can be purchased for $900 each, $700 less than presumably the open market price of $1,600, replacement now (losing one year on each existing machine) will buy four years of economic service. Balanced against an average cost of $2,048 for 6 years normal service, these replacement machines will have an average cost of:
   $900(A/P,20%,4) +[$480(P/F,20%,1) +$730(P/F,20%,2)
   +$1,050(P/F,20%,3) +$1,580(P/F,20%,4)](A/P,20%,4)
   = **$1,199**

115

PEE Solutions Manual   Chapter 11

11-15 and 11-16
Defender Analysis: Before income tax EUAC = $8,000(A/P,20%,10)
    +$38,400 +$11,520 +$325 +$1,000 +$900 = $8,000(0.23852)
    +$52,145 = $54,053
After income tax EUAC = ($8,000 +$8,000)(A/P,12%,10) +$52,145
    -0.4($52,145 +$2,000) = $16,000(0.17698) +$30,487 = $33,319

Challenger Analysis (Headings same as Problem 11-1)

0	($55,000)				($55,000)
1	($39,950)	($15,730)	($55,680)	$22,272	($17,678)
2	($39,950)	($11,220)	($51,170)	$20,468	($19,482)
3	($39,950)	($8,030)	($47,980)	$19,192	($20,758)
4	($39,950)	($5,720)	($45,670)	$18,268	($21,682)
5	($39,950)	($4,785)	($44,735)	$17,894	($22,056)
6	($39,950)	($4,785)	($44,735)	$17,894	($22,056)
7	($39,950)	($4,730)	($44,680)	$17,872	($22,078)
8	($39,950)		($39,950)	$15,980	($23,970)
9	($39,950)		($39,950)	$15,980	($23,970)
10	($39,950)		($39,950)	$15,980	($23,970)
Sums	($454,500)	($55,000)	($454,500)	$181,800	($272,700)
	($222,470) =NPV @20%			NPV @12%	($174,586)
	($53,064) =EUAC			EUAC =	($30,898)

Challenger is preferred both before and after income taxes.

PEE Solutions Manual   Chapter 11

11-17
(Headings same as Problem 11-1)

0	($ 55,000)			$ 5,500	($ 49,500)
1	($ 39,950)	($ 10,000)	($ 49,950)	$ 19,980	($ 19,970)
2	($ 39,950)	($ 9,000)	($ 48,950)	$ 19,580	($ 20,370)
3	($ 39,950)	($ 8,000)	($ 47,950)	$ 19,180	($ 20,770)
4	($ 39,950)	($ 7,000)	($ 46,950)	$ 18,780	($ 21,170)
5	($ 39,950)	($ 6,000)	($ 45,950)	$ 18,380	($ 21,570)
6	($ 39,950)	($ 5,000)	($ 44,950)	$ 17,980	($ 21,970)
7	($ 39,950)	($ 4,000)	($ 43,950)	$ 17,580	($ 22,370)
8	($ 39,950)	($ 3,000)	($ 42,950)	$ 17,180	($ 22,770)
9	($ 39,950)	($ 2,000)	($ 41,950)	$ 16,780	($ 23,170)
10	($ 39,950)	($ 1,000)	($ 40,950)	$ 16,380	($ 23,570)
Sums	($ 454,500)	($ 55,000)	($ 454,500)	$ 181,800	($ 267,200)

($222,469) =NPV @20%   NPV @12% ($ 170,429)
($53,069) =EUAC   EUAC = ($ 30,165)

PEE Solutions Manual   Chapter 11

11-18

The following cash flow tables for Plan A and Plan B are necessary to obtain the after-tax cash flows for each plan. The values of the after-tax cash flows are all negative. Therefore, the rate of return must be obtained by subtracting the year-by-year cash flows for Plan A from those for Plan B and computing the prospective rate of return on the additional investment required for B over A.

Plan A (Headings same as Problem 11-1)

0	($ 30,000)			($ 2,400)	($ 32,400)
1	($ 39,500)	($ 6,000)	($ 45,500)	$ 18,200	($ 21,300)
2	($ 39,500)	($ 5,040)	($ 44,540)	$ 17,816	($ 21,684)
3	($ 39,500)	($ 4,272)	($ 43,772)	$ 17,509	($ 21,991)
4	($ 39,500)	($ 3,360)	($ 42,860)	$ 17,144	($ 22,356)
5	($ 39,500)	($ 3,216)	($ 42,716)	$ 17,086	($ 22,414)
6	($ 39,500)	($ 2,823)	($ 42,323)	$ 16,929	($ 22,571)
7	($ 39,500)	($ 2,823)	($ 42,323)	$ 16,929	($ 22,571)
8	($ 39,500)	($ 2,823)	($ 42,323)	$ 16,929	($ 22,571)
9	($ 39,500)	($ 2,823)	($ 42,323)	$ 16,929	($ 22,571)
10	($ 39,500)	($ 2,823)	($ 42,323)	$ 14,129	($ 18,371)
10	$ 7,000				
Sums	($ 418,000)	($ 36,003)	($ 431,003)	$ 167,201	($ 250,799)

Plan B (Headings same as Problem 11-1)

0	($ 40,000)				($ 40,000)
1	($ 32,000)	($ 8,000)	($ 40,000)	$ 16,000	($ 16,000)
2	($ 32,000)	($ 6,400)	($ 38,400)	$ 15,360	($ 16,640)
3	($ 32,000)	($ 5,120)	($ 37,120)	$ 14,848	($ 17,152)
4	($ 32,000)	($ 3,600)	($ 35,600)	$ 14,240	($ 17,760)
5	($ 32,000)	($ 3,360)	($ 35,360)	$ 14,144	($ 17,856)
6	($ 32,000)	($ 2,720)	($ 34,720)	$ 13,888	($ 18,112)
7	($ 32,000)	($ 2,700)	($ 34,700)	$ 13,880	($ 18,120)
8	($ 32,000)	($ 2,700)	($ 34,700)	$ 13,880	($ 18,120)
9	($ 32,000)	($ 2,700)	($ 34,700)	$ 13,880	($ 18,120)
10	($ 32,000)	($ 2,700)	($ 19,700)	$ 7,880	($ 9,120)
10	$ 15,000				
Sums	($ 345,000)	($ 40,000)	($ 345,000)	$ 138,000	($ 207,000)

PEE Solutions Manual   Chapter 11

(a) Before-Tax Analysis   (b) After-tax Analysis

	Plan B	Plan A	(B-A)
0	($ 40,000)	($ 30,000)	($ 10,000)
1	($ 32,000)	($ 39,500)	$ 7,500
2	($ 32,000)	($ 39,500)	$ 7,500
3	($ 32,000)	($ 39,500)	$ 7,500
4	($ 32,000)	($ 39,500)	$ 7,500
5	($ 32,000)	($ 39,500)	$ 7,500
6	($ 32,000)	($ 39,500)	$ 7,500
7	($ 32,000)	($ 39,500)	$ 7,500
8	($ 32,000)	($ 39,500)	$ 7,500
9	($ 32,000)	($ 39,500)	$ 7,500
10	($ 32,000)	($ 39,500)	$ 15,500
10	$ 15,000	$ 7,000	
		IRR B/T =	74.9%

	Plan B	Plan A	(B-A)
0	($ 40,000)	($ 32,400)	($ 7,600)
1	($ 16,000)	($ 21,300)	$ 5,300
2	($ 16,640)	($ 21,684)	$ 5,044
3	($ 17,152)	($ 21,991)	$ 4,839
4	($ 17,760)	($ 22,356)	$ 4,596
5	($ 17,856)	($ 22,414)	$ 4,558
6	($ 18,112)	($ 22,571)	$ 4,459
7	($ 18,112)	($ 22,571)	$ 4,459
8	($ 18,112)	($ 22,571)	$ 4,459
9	($ 18,112)	($ 22,571)	$ 4,459
10	($ 9,120)	($ 18,371)	$ 9,251
		IRR A/T =	66.8%

(c) The before-tax rate of return of 74.9% on the extra $10,000 investment in Plan B over Plan A means that the two plans would be equivalent before income taxes with an interest rate of 74.9%. Similarly, the 66.8% incremental rate of return on the extra investment of $7,600 in B over A after income taxes is the rate of return that would make the two plans equivalent. In either case, Plan B is much preferred to Plan A.

11-19
(a) The total investment in purchase and installation of machine X = $120,000 +$24,000 = $144,000. The total savings for 8 years is: $32,000 +$25,000 +$22,000 +$19,000 +$16,000 +$13,000 +$10,000 +$7,000 = $144,000.
Therefore the rate of return before income taxes will be zero.
(b) This rate of return is not relevant to the decision at hand. The $120,000 is a sunk cost, and only the portion of it representing the current market value, $36,000 and the installation cost, $24,000, should be considered.
(c) If Alternative 1 is chosen there is no investment. If Alternative 2 is chosen there is an opportunity cost investment of the market value of Machine X, $36,000, plus its installation cost, $24,000, giving a total investment of $60,000.

PEE Solutions Manual   Chapter 11

End of Year	Cash flow Difference Alt (2-1)	End of Year	Cash flow Difference Alt (2-1)
1988=0	-$60,000	5	+16,000
1	+ 32,000	6	+13,000
2	+ 25,000	7	+10,000
3	+ 22,000	1996=8	+ 7,000
4	+ 19,000		

(d) Before-tax rate of return: Find i so that
$$NPW = 0 = -\$60,000 + (\$32,000 - \$28,000)(P/F,i\%,1) + \$28,000(P/A,i\%,8) - \$3,000(P/G,i\%,8)$$
$i = \underline{34.7\%}$ Choose Alternative 2.

(e) and (f) After-Tax Cash Flow Table (Headings same as Problem 11-1)

0	($60,000)	Loss =	$84,000*	($33,600)	($93,600)
1	$32,000	($41,184)	($9,184)	$3,674	$35,674
2	$25,000	($29,376)	($4,376)	$1,750	$26,750
3	$22,000	($21,024)	$976	($390)	$21,610
4	$19,000	($14,976)	$4,024	($1,610)	$17,390
5	$16,000	($12,485)	$3,515	($1,406)	$14,594
6	$13,000	($12,485)	$515	($206)	$12,794
7	$10,000	($12,470)	($2,470)	$988	$10,988
8	$7,000		$7,000	($2,800)	$4,200
SUMS	$84,000	($144,000)	$84,000	($33,600)	$50,400
				IIR A/T =	15.0%

Foregone loss on disposal = ($120,000 - $36,000) = $84,000. This solution also assumes that the $24,000 installation cost was capitalized and depreciated along with the $120,000 cost Machine X.

(g) Alternative 2 is preferred at any MARR of 15% or less. If the company's MARR is greater than 15%, alternative 1 is preferred.

120

PEE Solutions Manual   Chapter 11

11-20
(a) The economic life, n^*, is 6 years because:
AC_6 is minimum of all $EUAC_t$'s

At $t = 6$: $MC_6 = \$14,825 \leq AC_6 = \$15,143$

and $MC_7 = \$16,025 > AC_6$

(b)

(n)	t	MC_t	P/F	ΣPW	A/P	AC_t
2	1	$13,400	0.8696	$11,653	1.15	$13,400
3	2	12,950	0.7561	21,444	0.61512	13,191
4	3	13,000	0.6575	29,992	0.4380	_13,136_
5	4	13,625	0.5718	37,783	0.3503	13,234

DAC = $13,136 = AC_3 (That is, $MC_4 = \$13,625 > AC_3 = \$13,136$)

CAC = $AC_6 = \$15,143$ (as before).

Therefore, DO NOT REPLACE NOW
Note that _implied_ remaining life is 3 yrs for a 1 yr-old asset. Thus total life would be 4 yrs, not 6 yrs as previously found. THIS CAN'T BE. Remaining life has _no meaning_ in this context.

(c) <u>2 yrs old</u>

(n)	t	MC_t
3	1	$12,950
4	2	$13,000

DAC = $MC_1 = \$12,950$

CAC = $15,143   Therefore, DO NOT REPLACE NOW.
  Note from the original table that, from $t = 4$ on, the marginal costs are increasing constantly. Therefore, from age 3 on the DAC always will be the next year MC.

<u>5 yrs old</u>

(n)	t	MC_t
6	1	$14,825
7	2	16,025

; Thus DAC = $MC_1 = \$14,825$

CAC = $15,143   Therefore, DO NOT REPLACE NOW.

<u>6 yrs old</u>

(n)	t	MC_t
7	1	$16,025
8	2	16,300
9	3	16,650
10	4	16,800

; Thus DAC = $MC_1 = \$16,025$

CAC = $15,143   REPLACE NOW

11-20(cont)
Curves of Equivalent Uniform Annual Costs and Marginal Costs for Problem 11-20, (a)-(c)

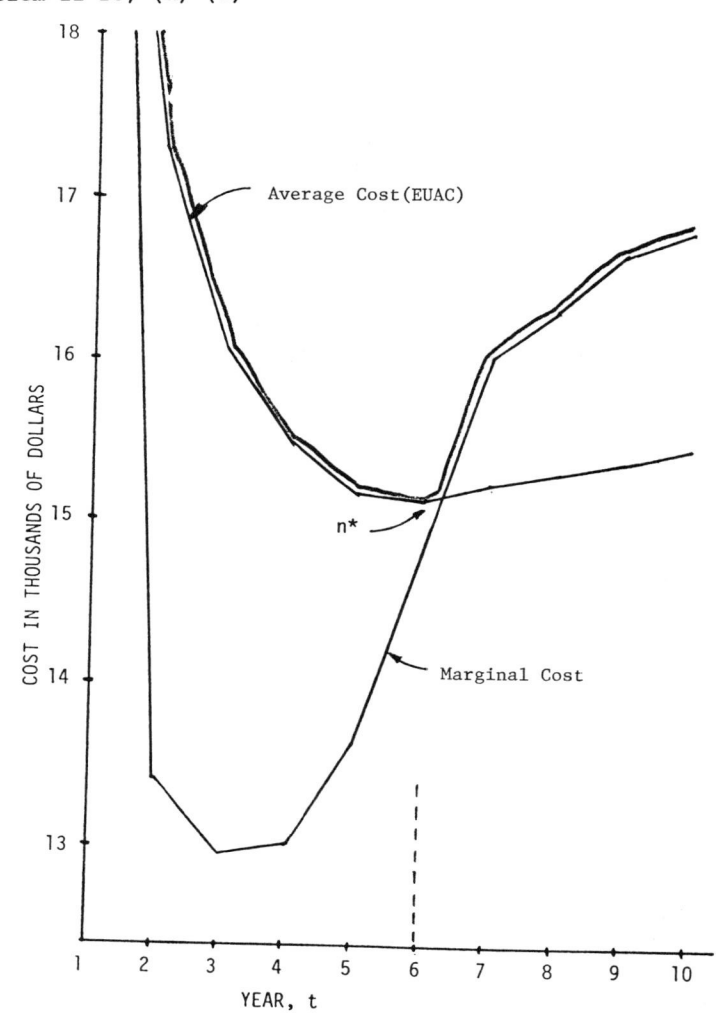

PEE Solutions Manual   Chapter 11

11-20(cont)
 (d) Treat defender salvage value (F_6 = $3,500) as a reduction in FC of Challenger and current defender age as n for challenger.

Defender DAC = A_7 = **$14,000**

Challenger CAC = AC_6 = [($25,000 -$3,500) -$3,500](A/P,15%,6)
+$3,500(0.15) +(A/P,15%,6)[$8,000(P/A,15%,3)
+$9,000(P/F,15%,4) +$10,500(P/F,15%,5) +$12,000(P/F,15%,6)]
= $4,756 +$525 +$8,936 = **$14,217**

 Solution says DO NOT REPLACE. The problem here is that the $3,500 salvage value of the defender in year 6 is subtracted from the purchase price of the challenger and, in effect, is amortized over the 6 year life of the challenger rather than over its (correct) 3 year remaining life. This unjustly penalizes the challenger.

11-21
 Defender Analysis (Headings same as Problem 11-1)

0	($12,000)				($12,000)
1	($25,000)	($6,500)	($31,500)	$12,600	($12,400)
2	($28,000)	($6,500)	($34,500)	$13,800	($14,200)
3	($31,000)	($6,500)	($37,500)	$15,000	($16,000)
4				NPV =	($44,040)
5				EUAC =	($19,288)

Challenger Analysis

0	($55,000)				($55,000)
1	($16,000)	($22,000)	($38,000)	$15,200	($800)
2	($18,500)	($13,200)	($31,700)	$12,680	($5,820)
3	($21,000)	($7,920)	($28,920)	$11,568	($9,432)
4	($23,500)	($5,940)	($29,440)	$11,776	($11,724)
5	($26,000)	($5,940)	($31,940)	$12,776	($13,224)
6	($28,500)		($20,500)	$8,200	($12,300)
6	$8,000				
				NPV =	($84,894)
				EUAC =	($22,432)

Replacement should not be made now.

PEE Solutions Manual   Chapter 11

11-22
Economic Life Determination

	Value	Decrease	Interest	Disburse-	Marginal	Sum PW for	EUAC if
t end of yr		in value	on beg. val	ments	cost yr t	t years	kept t yrs
0	$55,000						
1	$37,000	$18,000	$13,750	$16,000	$47,750	$38,200	$47,750
2	27,000	10,000	9,250	18,500	37,750	62,360	43,305
3	20,000	7,000	6,750	21,000	34,750	80,152	41,062
4	15,000	5,000	5,000	23,500	33,500	93,874	39,750
5	11,000	4,000	3,750	26,000	33,750	104,933	39,020
6	8,000	3,000	2,750	28,500	34,250	113,910	38,595
7	6,000	2,000	2,000	31,000	35,000	121,250	38,356
8	4,000	2,000	1,500	33,500	37,000	127,459	38,229

11-23

	Salvage Value end of year	Decrease in value	Interest on Beg. of year value	Disburse- ments	Marginal cost of that year	PW @ 30%	Sum of PW for n years	EUAC if sold at end of year
0	($ 32,000)							
1	($ 26,000)	($ 6,000)	($ 9,600)	($ 18,000)	($ 33,600)	($ 25,845)	($ 25,845)	($ 35,599)
2	($ 20,000)	($ 6,000)	($ 7,800)	($ 19,500)	($ 33,300)	($ 19,704)	($ 45,549)	($ 33,469)
3	($ 16,000)	($ 4,000)	($ 6,000)	($ 21,000)	($ 31,000)	($ 14,111)	($ 59,660)	($ 32,850)
4	($ 13,000)	($ 3,000)	($ 4,800)	($ 22,500)	($ 30,300)	($ 10,608)	($ 70,268)	($ 32,438)
5	($ 11,000)	($ 2,000)	($ 3,900)	($ 24,000)	($ 29,900)	($ 8,052)	($ 78,320)	($ 32,157)
6	($ 9,000)	($ 2,000)	($ 3,300)	($ 25,500)	($ 30,800)	($ 6,382)	($ 84,702)	($ 32,051)
7	($ 7,000)	($ 2,000)	($ 2,700)	($ 27,000)	($ 31,700)	($ 5,053)	($ 89,755)	($ 32,031)
8	($ 5,000)	($ 2,000)	($ 2,100)	($ 28,500)	($ 32,600)	($ 3,997)	($ 93,751)	($ 32,055)
9	($ 4,000)	($ 1,000)	($ 1,500)	($ 30,000)	($ 32,500)	($ 3,065)	($ 96,816)	($ 32,069)
10	$ 0		($ 1,200)	($ 31,500)	($ 32,700)	($ 2,371)	($ 99,187)	($ 32,083)

The apparent economic life is 7 years.

PEE Solutions Manual   Chapter 11

11-24

After-tax analysis of Problem 11-23

	Cash flow before income tax	Write-off for Depreciat'n	Taxable Income	Cash flow for income taxes	Cash flow after income tax	PW @ 15%	sum of PWs if sold at end of yr.	EUAC if sold at end of year
0	($ 32,000)				($ 32,000)	($ 32,000)		
1	($ 18,000)	($ 3,200)	($ 21,200)	$ 8,480	($ 9,520)	($ 8,279)		
	$ 26,000	If sold loss:no cap.gains tax			$ 26,000	$ 22,610	($ 17,669)	($ 20,319)
2	($ 19,500)	($ 3,200)	($ 22,700)	$ 9,080	($ 10,420)	($ 7,879)		
	$ 20,000	If sold loss: no cap. gains tax			$ 20,000	$ 15,122	($ 33,035)	($ 20,320)
3	($ 21,000)	($ 3,200)	($ 24,200)	$ 9,680	($ 11,320)	($ 7,443)		
	$ 16,000	If sold loss: no cap. gains tax			$ 16,000	$ 10,520	($ 45,080)	($ 19,748)
4	($ 22,500)	($ 3,200)	($ 25,700)	$ 10,280	($ 12,220)	($ 6,987)		
	$ 13,000	If sold loss: no cap. gains tax			$ 13,000	$ 7,433	($ 55,154)	($ 19,319)
5	($ 24,000)	($ 3,200)	($ 27,200)	$ 10,880	($ 13,120)	($ 6,523)		
	$ 11,000	If sold loss: no cap. gains tax			$ 11,000	$ 5,469	($ 63,642)	($ 18,985)
6	($ 25,500)	($ 3,200)	($ 28,700)	$ 11,480	($ 14,020)	($ 6,061)		
	$ 9,000	If sold loss: no cap. gains tax			$ 9,000	$ 3,891	($ 71,201)	($ 18,833)
7	($ 27,000)	($ 3,200)	($ 30,200)	$ 12,080	($ 14,920)	($ 5,608)		
	$ 7,000	If sold loss: no cap. gains tax			$ 7,000	$ 2,631	($ 78,149)	($ 18,784)
8	($ 28,500)	($ 3,200)	($ 31,700)	$ 12,680	($ 15,820)	($ 5,172)		
	$ 5,000	If sold loss: no cap. gains tax			$ 5,000	$ 1,635	($ 84,317)	($ 18,790)
9	($ 30,000)	($ 3,200)	($ 33,200)	$ 13,280	($ 16,720)	($ 4,753)		
	$ 4,000	If sold, gain	$800	($ 320)	$ 3,680	$ 1,046	($ 89,659)	($ 18,790)
10	($ 31,500)	($ 3,200)	($ 34,700)	$ 13,880	($ 17,620)	($ 4,356)	($ 95,061)	($ 18,941)

11-25
(Headings same as Problem 11-1)

0	($ 40,000)				($ 40,000)
1	($ 14,000)	($ 11,440)	($ 25,440)	$ 10,176	($ 3,824)
2	($ 16,000)	($ 8,160)	($ 24,160)	$ 9,664	($ 6,336)
3	($ 18,000)	($ 5,840)	($ 23,840)	$ 9,536	($ 8,464)
4	($ 20,000)	($ 4,160)	($ 24,160)	$ 9,664	($ 10,336)
5	($ 22,000)	($ 3,467)	($ 25,467)	$ 10,187	($ 11,813)
6	($ 24,000)	($ 3,467)	($ 27,467)	$ 10,987	($ 13,013)
7	($ 26,000)	($ 3,466)	($ 29,466)	$ 11,786	($ 14,214)
8	($ 28,000)		($ 28,000)	$ 11,200	($ 16,800)
				NPV =	($ 71,240)
				EUAC =	($ 15,876)

PEE Solutions Manual

CHAPTER 12

Financing Effects on Economy Studies

General Notes:
A great many mistakes are made in business when the prospective rate of return on just the equity portion of a proposed investment is used as the measure of attractiveness. The decision should be based on the prospective rate of return on a proposed project as if it were to be entirely financed from equity funds. That is the only true measure of the attractiveness of an investment. The means of financing is a separable decision and should be kept separate from the evaluation of a project.

When a portion of an investment is to be financed by borrowed money at an interest rate lower than the i earned by the project, the RoR on equity will be higher than i. As the ratio of borrowed to equity funds increases, the RoR on equity will approach infinity as the equity portion decreases to zero. Obviously, measuring the attractiveness of a proposal by RoR on equity is like using a rubber tape measure. It is not reliable.

A second major topic in Chapter 12 is the matter of leasing versus ownership. This chapter clearly shows that leasing or renting are forms of borrowing. A person who realizes that costs of leasing and borrowing can be measured in the same way is in a favorable position to evaluate the sales arguments of leasing agents. Leasing is justified in many situations, but the decision to lease is too frequently made for the wrong reasons.

The teacher can bring the student to understand leasing by using examples, such as the leasing of private autos, for which the student can obtain information from local agents. The experience will usually make leasing more understandable and "real" to the student.

By this point in the course, the student should be able to handle some rather difficult analysis problems. This chapter gives the teacher an opportunity to find out how well the student has learned basic principles.

12-1

End of Year	Plan P	Plan R	Plan P - Plan R
0	-100N	-10N - 24N	-66N
1		-24N	+24N
2		-24N	+24N
3	+ 55N	+10N	+45N

The interest rate that equates the two plans is found from:
PW = 0 = -66N + 24N(P/A,i%,2) + 45N(P/F,i%,3); i = <u>17.1%</u>

PEE Solutions Manual   Chapter 12

12-2
Plan P after-tax

Year	Cash flow before income taxes	Write-off of initial outlay for tax purposes	Influence on taxable income	Effects of taxes on cash flow	Cash flow after income taxes
0	($100.00)				($100.00)
1		($40.00)	($40.00)	$16.00	$16.00
2		($24.00)	($24.00)	$9.60	$9.60
3		($14.40)	($14.00)	$5.60	$47.40
3	$55.00, Cap. gains =		$33.00	($13.20)	
	B.V.(3) =	$22.00		SUM =	($27.00)

Plan R after-tax

0	($34.00)				($34.00)
1	($24.00)		($24.00)	$9.60	($14.40)
2	($24.00)		($24.00)	$9.60	($14.40)
3	$10.00		($24.00)	$9.60	$19.60

	Plan P	Plan R	(P - R)	
0	($100,00)	($34.00)	($66.00)	Cost of "borrowed" money
1	$16.00	($14.40)	$30.40	by leasing = 11.4% after
2	$9.00	($14.40)	$23.40	income tax.
3	$47.00	$19.60	$27.40	

12-3
Consider first the before-tax analysis in the solution to Problem 12-1. Cash flow for Plan R will be unchanged whereas the positive cash flow for Plan P in year 3 will be reduced below 55N. Therefore the difference in cash flow in year 3 will be less than the +45N indicated in the solution. The before-tax cost of money will be reduced. The expectation of a lower salvage value makes leasing more attractive to a prospective lessee. This answer assumes that the lessor's price is not influenced by the lessee's expectation about salvage value.

The foregoing statement also applies to an after-tax analysis provided the lessee's estimate does not influence either the allowable depreciation deduction from taxable income or the required payments under the lease. If one or both of these will be affected, a general answer is not possible; each case calls for its own analysis.

12-4

The before-tax cost of the money provided by the lessor in Problem 12-1 was 17.1%; if the applicable tax rate had been 0%, this would also have been the after-tax cost of the money. With an applicable tax rate of 40% in Problem 12-2, this after-tax cost was reduced to 11.4%. If the tax rate had been 50% in Problem 12-2, the cost would have been 9.1%. In general, high applicable income tax rates tend to encourage financing by leasing rather than ownership just as they tend to encourage financing by borrowing rather than ownership. The extent of encouragement depends on the rules established by the taxing authorities about rates of write-off if assets are owned and about the deductibility of payments to the lessor if assets are leased.

12-5

(a) 0 = -$60,000 + $19,400(P/A,i%,5)
     i = approximately <u>18.5%</u>

(b) Debt charge the first year will be $10,000 + (0.12)($50,000) = $16,000. This will decrease by (0.12)($10,000) = $1,200 each year. The positive cash flow in year 1 will be $19,400 - $16,000 = $3,400. This will increase by $1,200 each year.

    0 = -$10,000 + $3,400(P/A,i%,5) + $1,200(P/G,i%,5)

    i = approximately <u>41.3%</u>

PEE Solutions Manual   Chapter 12

12-6
After-tax analysis with equity financing
(Headings same as Problem 12-2)

0	($ 60,000)				($ 60,000)
1	$ 19,400	($ 24,000)	($ 4,600)	$ 1,840	$ 21,240
2	$ 19,400	($ 14,400)	$ 5,000	($ 2,000)	$ 17,400
3	$ 19,400	($ 8,640)	$ 10,760	($ 4,304)	$ 15,096
4	$ 19,400	($ 6,480)	$ 12,920	($ 5,168)	$ 14,232
5	$ 19,400	($ 6,480)	$ 12,920	($ 5,168)	$ 14,232
				IRR A/T =	12.5%

After-tax analysis with debt financing

year	Cash flow before income taxes	Write-off of initial outlay for Tax purposes	Payment on debt	Payment of interest	Influence on taxable income	Influence of income taxes on cash flow	Cash flow after income taxes
0	($ 60,000)						
0	$ 50,000 loan @12%						($ 10,000)
1	$ 19,400	($ 24,000)	($ 10,000)	($ 6,000)	($ 10,600)	$ 4,240	$ 7,640
2	$ 19,400	($ 14,400)	($ 10,000)	($ 4,800)	$ 200	($ 80)	$ 4,520
3	$ 19,400	($ 8,640)	($ 10,000)	($ 3,600)	$ 7,160	($ 2,864)	$ 2,936
4	$ 19,400	($ 6,480)	($ 10,000)	($ 2,400)	$ 10,520	($ 4,208)	$ 2,792
5	$ 19,400	($ 6,480)	($ 10,000)	($ 1,200)	$ 11,720	($ 4,688)	$ 3,512
						IRR A/T =	41.6%

12-7 & 12-8

Before tax rate of return is found from:
NPW = 0 = -$27,000 +$7,440(P/A,i%,9) +$840(P/G,i%,9); <u>i = 31.9%</u>

130

PEE Solutions Manual   Chapter 12

12-7 & 12-8 (cont)
Equity Financing (Headings same as Problem 12-6)

Year						
0	($ 90000)					($ 90000)
1	$ 22000	($ 25714)	($ 3714)		$ 1486	$ 23486
2	$ 22000	($ 18366)	$ 3634		($ 1454)	$ 20546
3	$ 22000	($ 13118)	$ 8882		($ 3553)	$ 18447
4	$ 22000	($ 9378)	$ 12622		($ 5049)	$ 16951
5	$ 22000	($ 7808)	$ 14192		($ 5677)	$ 16323
6	$ 22000	($ 7808)	$ 14192		($ 5677)	$ 16323
7	$ 22000	($ 7808)	$ 14192		($ 5677)	$ 16323
8	$ 22000		$ 22000		($ 8800)	$ 13200
9	$ 22000		$ 22000		($ 8800)	$ 13200
Totals	108000	($ 90000)			($ 43200)	$ 64800

IRR = 19.5% Before tax   IRR After Tax = 14.0%

Debt Financing

year	Cash flow before income taxes	Write-off of initial outlay for Tax purposes	Payment on debt	Payment of interest	Influence on taxable income	Influence of income taxes on cash flow -0.40F	Cash flow after income taxes
0	($ 90,000)						
0	$ 63,000						($ 27,000)
1	$ 22,000	($ 25,714)	($ 7,000)	($ 7,560)	($ 11,274)	$ 4,510	$ 11,950
2	$ 22,000	($ 18,366)	($ 7,000)	($ 6,720)	($ 3,086)	$ 1,234	$ 9,514
3	$ 22,000	($ 13,118)	($ 7,000)	($ 5,880)	$ 3,002	($ 1,201)	$ 7,919
4	$ 22,000	($ 9,378)	($ 7,000)	($ 5,040)	$ 7,582	($ 3,033)	$ 6,927
5	$ 22,000	($ 7,808)	($ 7,000)	($ 4,200)	$ 9,992	($ 3,997)	$ 6,803
6	$ 22,000	($ 7,808)	($ 7,000)	($ 3,360)	$ 10,832	($ 4,333)	$ 7,307
7	$ 22,000	($ 7,808)	($ 7,000)	($ 2,520)	$ 11,672	($ 4,669)	$ 7,811
8	$ 22,000		($ 7,000)	($ 1,680)	$ 20,320	($ 8,128)	$ 5,192
9	$ 22,000		($ 7,000)	($ 840)	$ 21,160	($ 8,464)	$ 5,696
Totals	$ 171,000	($ 90,000)	($ 63,000)	($ 37,800)	$ 70,200	($ 28,080)	$ 42,120

Irr After Tax = 29.4%

12-9
Assume the first cost of the hydroelectric power project is P and the life is n (less than 50 years). Then the costs the magazine writer would include are:
- (1) P(i)
- (2) P/n
- (3) P(A/F,i%,n)
- (4) P(A/P,i%,n) or its non-uniform equivalent
- (5) Operation and maintenance costs

Thus capital recovery is provided for thrice, and interest on capital recovery twice. Either (1) and (3), or (4) alone, would be adequate to provide for capital recovery with interest, since:
P(i) + P(A/F,i%,n) = P(A/P,i%,n)
(2) alone would provide for just capital recovery without recognizing the time value of money.

Operation and maintenance costs (5) should be included with any <u>one</u> of the above provisions for capital recovery with interest to arrive at total annual cost.

12-10
Total expenses, as tabulated from the garage owner's association report, are $445,950 per year.

Calculated at 7% interest, the EUAC to the city is ($3,200,000 - $700,000)(A/P,7%,50) +$700,000(0.07) = $181,150 +$49,000 = $230,150. This is less than half the cost quoted above.

1. The depreciation charge assumes recovery of the full cost of the construction of the structure over 50 years at 0% interest. (A/P,0%,50) = 1/50.

2. Bond recovery with interest assumes payment of all costs, building and land, over 20 years. The interest charge of $224,000 is the first year charge only. It will decline by 0.07($160,000) = $11,200 annually after the first year.

3. The sinking fund at 5% provides for recovery of the full capital investment over 50 years. This all adds up to triple counting of various investment costs over the period.

It has been assumed that the structure will have 0 salvage at the end of 50 years. If revenues were set such that the debt obligation of the city were covered ($160,000 +$224,000 = $384,000), the garage would be self-sustaining and would more than pay for itself. Another option, to reduce annual costs, would be to lease air rights above the garage.

12-11

(a) Because of an improper choice of alternatives to compare, Mr. A's analysis is misleading. However, he has correctly computed the differences in cash flow after taxes for the lease vs. purchase situation. In addition, his calculation of the compound amount ten years hence is correct, assuming that funds made available through leasing will earn 20% net after taxes in other available alternative uses on a risk-adjusted basis. This may be verified as follows: The PW of the difference in cash flow after taxes at 20% is:

$100,000 −$25,000(P/A,20%,5) +$1,000(P/G,20%,5)
−[$8,000(P/A,20%,4) −$1,000(P/G,20%,4)](P/F,20%,5)
+$6,000(P/F,20%,10) = $24,106

Then the compound amount in 10 years is $24,106(F/P,20%,10) = $149,265, substantially the same as Mr. A's $149,300.

(b) The before-tax cost of money obtained from leasing:
0 = +$100,000 −$30,000(P/A,i%,5) −$6,000(P/A,i%,4)(P/F,i%,5)
+$4,000(P/F,i%,10); <u>i = 18.2%</u>

(c) The after-tax cost of money obtained from leasing:
0 = +$100,000 −$25,000(P/A,i%,5) +$1,000(P/G,i%,5)
−[$8,000(P/A,i%,4) −$1,000(P/G,i%,4)](P/F,i%,5)
+$6,000(P/F,i%,10); <u>i = 9.5%</u>

(d) <u>It is not relevant to use the compound amount at some future date as a decision criterion concerning present decisions.</u>
Mr. A. is assuming that, if the $100,000 were not available from leasing to earn the 20%, these earnings would be foregone, that is, that there are no other sources of investment funds available to the company. The analysis would be properly structured, however, if he considered either:

    (i) the difference between earnings from funds made available from leasing and funds made available from a bank line of credit or other source, since other sources are normally available; or

    (ii) omitted completely the consideration of working capital earnings, since they would be the same regardless of their source. What does vary is the cost of money from its various sources.

It would be better if he chose (ii), but if he feels he must give weight to what happens to the freed funds, he may choose (i). Procedure (ii) is usually followed in economy studies to simplify the analysis.

Leasing would be preferable if the after-tax cost of money from leasing, 9-1/2%, were less than the cost of funds available from other sources after taxes (which it probably is not). Thus it would be necessary to know either the cost of other funds or some statement of the after-tax minimum attractive rate of return for projects of this nature.

PEE Solutions Manual   Chapter 12

12-12
Equity financing (Headings same as Problem 12-6)

0	($ 100.00)				($ 100.00)
1		($ 20.00)	($ 20.00)	$ 8.00	$ 8.00
2		($ 16.00)	($ 16.00)	$ 6.40	$ 6.40
3		($ 12.80)	($ 12.80)	$ 5.12	$ 5.12
4		($ 9.00)	($ 9.00)	$ 3.60	$ 3.60
5		($ 8.40)	($ 8.40)	$ 3.36	$ 3.36
6		($ 6.76)	($ 6.76)	$ 2.70	$ 2.70
7		($ 6.75)	($ 6.75)	$ 2.70	$ 2.70
8		($ 6.75)	($ 6.75)	$ 2.70	$ 2.70
9		($ 6.75)	($ 6.75)	$ 2.70	$ 2.70
10		($ 6.75)	($ 6.75)	$ 2.70	$ 2.70

Leasing (Headings same as Problem 12-6)

						(L - E)
0	($ 38.00)			($ 38.00)	$ 62.00	
1	($ 20.00)	($ 20.00)	$ 8.00	($ 12.00)	($ 20.00)	
2	($ 20.00)	($ 20.00)	$ 8.00	($ 12.00)	($ 18.40)	
3	($ 20.00)	($ 20.00)	$ 8.00	($ 12.00)	($ 17.12)	
4	($ 16.00)	($ 20.00)	$ 8.00	($ 8.00)	($ 11.60)	
5	($ 16.00)	($ 16.00)	$ 6.40	($ 9.60)	($ 12.96)	
6	($ 16.00)	($ 16.00)	$ 6.40	($ 9.60)	($ 12.30)	
7	($ 10.00)	($ 16.00)	$ 6.40	($ 3.60)	($ 6.30)	
8	($ 10.00)	($ 10.00)	$ 4.00	($ 6.00)	($ 8.70)	
9	($ 10.00)	($ 10.00)	$ 4.00	($ 6.00)	($ 8.70)	
10	$ 18.00	($ 10.00)	$ 4.00	$ 22.00	$ 19.30	
				NPV@16.2%	$ 0.00	

(Although there are two reversals of sign in the series that shows the prospective after-tax differences in cash flow, this is one of the types of case discussed in Appendix B where it is good enough for practical purposes to use conventional methods of analysis.)

12-13

With a given set of terms of a lease, the prospect of inflation makes leasing less attractive to the lessor for the same reason that it becomes more attractive to the lessee. As inflation continues, the manufacturer presumably will make annual rental charges a larger fraction of the first cost of any machine. Leasing prices tend to rise during periods of inflation for the same reason that interest rates tend to rise during such periods.

12-14

(a) The following disbursements tend to increase the city's equity in the water system:

New fixed assets:	New well	$13,300
	Pump house and pump	9,600
	New water mains	12,800
Reduction of liability:	Repayment of bonds	25,000
		$60,700

The foregoing outlays may reasonably be viewed as "financial provision for depreciation." Their total is higher than the city engineer's estimate of $45,270 for the depreciation chargeable in 1988.

(b) The city engineer's estimate of the net present value of the plant on January 1, 1988 was $619,235. This is greater than the outstanding debt of $400,000. If the city engineer's estimate of net present value is accepted, the city had an equity of $219,235 in its water plant at the start of 1988. According to the problem statement this equity started at zero when the plant was constructed and has been built up entirely by revenues from the sale of water. Therefore, it appears that the overall financial provisions for depreciation up to the start of 1988 were more than adequate.

12-15

The following three tables show the cash flow for RG Industries under the three conditions, 100% financing with equity capital, purchase with debt financing of $60,000 and financing through leasing. The rate of return before income taxes for equity financing is a little over 20% and the after-tax rate of return is 14.3%. Both of these rates exceed the criteria established by the company and, therefore, the project would be attractive if the company had enough available capital.

It is known that the cost of money borrowed from the bank will be 12% before taxes or about 7.2% after-tax. The cost of money through leasing is 12.3% after taxes. Therefore, the company will be better off to borrow the money for the machine. The after-tax cash flow for both borrowing and leasing show that the company can meet the payments if the project results are reasonably close to the estimates.

(a) Equity financing (Headings same as in Problem 12-6)

0	($ 80,000)				($ 80,000)
1	$ 18,000	($ 16,000)	$ 2,000	($ 800)	$ 17,200
2	$ 18,000	($ 12,800)	$ 5,200	($ 2,080)	$ 15,920
3	$ 18,000	($ 10,240)	$ 7,760	($ 3,104)	$ 14,896
4	$ 18,000	($ 7,200)	$ 10,800	($ 4,320)	$ 13,680
5	$ 18,000	($ 6,752)	$ 11,248	($ 4,499)	$ 13,501
6	$ 18,000	($ 5,408)	$ 12,592	($ 5,037)	$ 12,963
7	$ 18,000	($ 5,400)	$ 12,600	($ 5,040)	$ 12,960
8	$ 18,000	($ 5,400)	$ 12,600	($ 5,040)	$ 12,960
9	$ 18,000	($ 5,400)	$ 12,600	($ 5,040)	$ 12,960
10	$ 18,000	($ 5,400)	$ 12,600	($ 5,040)	$ 12,960
11	$ 18,000		$ 18,000	($ 7,200)	$ 10,800
12	$ 18,000		$ 18,000	($ 7,200)	$ 13,800
12	$ 5,000		$ 5,000	($ 2,000)	
		($ 80,000)		IRRA/T =	14.3%

PEE Solutions Manual   Chapter 12

(b) Debt financing (Headings same as Problem 12-2.)

0	($80,000)						
0	$60,000 Borrowed						($20,000)
1	$18,000	($16,000)	($7,500)	($7,200)	($5,200)	$2,080	$5,380
2	$18,000	($12,800)	($7,500)	($6,300)	($1,100)	$440	$4,640
3	$18,000	($10,240)	($7,500)	($5,400)	$2,360	($944)	$4,156
4	$18,000	($7,200)	($7,500)	($4,500)	$6,300	($2,520)	$3,480
5	$18,000	($6,752)	($7,500)	($3,600)	$7,648	($3,059)	$3,841
6	$18,000	($5,408)	($7,500)	($2,700)	$9,892	($3,957)	$3,843
7	$18,000	($5,400)	($7,500)	($1,800)	$10,800	($4,320)	$4,380
8	$18,000	($5,400)	($7,500)	($900)	$11,700	($4,680)	$4,920
9	$18,000	($5,400)			$12,600	($5,040)	$12,960
10	$18,000	($5,400)			$12,600	($5,040)	$12,960
11	$18,000				$18,000	($7,200)	$10,800
12	$18,000				$23,000	($9,200)	$8,800
12	$5,000			RoR on equity After tax=			24.4%

After-Tax Analysis of leasing

	Lease Payments	Taxable Income	CF for Taxes	CF after Taxes	(L - Equity)
0				($20,000)	$60,000
1	$18,000	($12,000)	$6,000	($2,400)	$3,600	($13,600)
2	$18,000	($12,000)	$6,000	($2,400)	$3,600	($12,320)
3	$18,000	($12,000)	$6,000	($2,400)	$3,600	($11,296)
4	$18,000	($12,000)	$6,000	($2,400)	$3,600	($10,080)
5	$18,000	($12,000)	$6,000	($2,400)	$3,600	($9,901)
6	$18,000	($12,000)	$6,000	($2,400)	$3,600	($9,363)
7	$18,000	($12,000)	$6,000	($2,400)	$3,600	($9,360)
8	$18,000	($12,000)	$6,000	($2,400)	$3,600	($9,360)
9	$18,000	($12,000)	$6,000	($2,400)	$3,600	($9,360)
10	$18,000	($12,000)	$6,000	($2,400)	$3,600	($9,360)
11	$18,000	($12,000)	$6,000	($2,400)	$3,600	($7,200)
12	$18,000	$8,000	$6,000	($2,400)	$23,600	$9,800

Cost of money = 12.3%

PEE Solutions Manual   Chapter 12

12-16

Year	Disburse-ments	Receipts	Write-off Depreciat'n	Taxable Income	C.F. for Taxes	C.F. After Taxes
0	($ 65,000)	$ 20,000				($ 45,000)
1		$ 12,000	($ 13,000)	($ 1,000)	$ 400	$ 12,400
2		$ 12,000	($ 10,400)	$ 1,600	($ 640)	$ 11,360
3		$ 12,000	($ 8,320)	$ 3,680	($ 1,472)	$ 10,528
4		$ 12,000	($ 5,850)	$ 6,150	($ 2,460)	$ 9,540
5		$ 12,000	($ 5,486)	$ 6,514	($ 2,606)	$ 9,394
6		$ 12,000	($ 4,394)	$ 7,606	($ 3,042)	$ 8,958
7		$ 12,000	($ 4,388)	$ 7,613	($ 3,045)	$ 8,955
8		$ 12,000	($ 4,388)	$ 7,613	($ 3,045)	$ 8,955
9		$ 12,000	($ 4,388)	$ 7,613	($ 3,045)	$ 8,955
10		$ 12,000	($ 4,388)	$ 7,613	($ 3,045)	$ 8,955
11		$ 12,000		$ 12,000	($ 4,800)	$ 7,200
12	($ 8,000)	$ 15,000		$ 24,000	($ 9,600)	($ 5,600)
12	($ 3,000)					

IRR A/T = 19.2%

12-17
This problem statement implies that an after tax analysis can be made for Bob, but to do so you would need the interest rate that he is paying on his loan of $18,000. Since he is earning about 12% on his investments before tax, then a before tax cost of borrowed money or of money made available through leasing is enough to give him a basis for decision.
Furthermore, his payments will be monthly and the depreciation allowances are stated as annual percentages. These would have to be broken down into monthly amounts, just adding to the difficulty of the problem without helping the students. The spreadsheet results show that his before-tax cost of money by the loan is about 19.4%, which is substantially greater than the 12% he is currently earning on his securities. In addition, the analysis shows that the cost of money by leasing is actually negative. Based on this analysis, he will definitely be better off financially if he leases.
This is a very unusual problem, because the cost of money by leasing is usually higher than bank loans. In this case, it seems the leasing company is "stuck" with a number of cars that it needs to lease, with the result that it has cut the rental charge a great deal. In problem 12-18, you find that the leasing company is still earning about 21.4% before-tax on its investment in the car.

138

PEE Solutions Manual   Chapter 12

	C.F. for	C.F. with	C.F. by	(Loan -	(Lease -
month	Purchase	car loan	leasing	Equity)	Equity)
0	($19,995)	($1,995)	($416)	$18,000	$19,579
1		($650)	($416)	($650)	($416)
2 to 35		($650)	($416)	($650)	($416)
36	$5,000	$4,350		($650)	($5,000)

Before-tax Effective interest rate =   19.4%   negative

12-18

Year	Cash flow before income taxes A	Write-off of initial outlay for tax purposes B	Influence on taxable income C	Influence of income taxes on cash flow -0.40C D	Cash flow after income taxes (A + D) E
0	-15910				-15910
1	4992	-6364	-1372	548.8	5540.8
2	4992	-3818.4	1173.6	-469.44	4522.56
3	4992	-2291.04	2700.96	-1080.384	3911.616
3	6000		2563.44	-1025.376	4974.624
Sums	5066	-12473.44	5066	-2026.4	3039.6

After tax IRR =   7.59%

PEE Solutions Manual   Chapter 12

12-19

A complete after-tax analysis of the bond proposal on the after-tax cash flow for the firm is most easily performed using a personal computer spreadsheet such as we have been using. Portions of such a spreadsheet follow. It shows that the firm will have a positive, after-tax, cash flow of $572,500 at the end of the first six-month period and that the positive cash flow will increase steadily throughout the 20 years (40 periods) if the estimated investment effects are accurate. The problem stated that the $50 million investment would increase the after-tax earnings by about $7.5 million a year. Since the income tax rate is 40%, that means that the before income tax earnings would be about ($7.5/0.6) = $12.5 million. That is used as the before-tax extra income produced as a result of the investment. It seems the company is fully justified in selling the bonds.

Semiannual payments and receipts

6 mos. period	Extra Income Before-Tax ($7.5/.6)	Write-off of fees & Commis'n	Bonds outstanding	Payment to redeem bonds	Payment of interest & fees	Taxable income	C.F. for Taxes	C.F. after taxes
0			($52,500.0)					
1	$6,250.0	($62.5)	($51,187.5)	($1,312.5)	($3,150.0)	$3,037.5	($1,215.0)	$572.5
2	$6,250.0	($62.5)	($49,875.0)	($1,312.5)	($3,071.2)	$3,116.3	($1,246.5)	$619.8
3	$6,250.0	($62.5)	($48,562.5)	($1,312.5)	($2,992.4)	$3,195.1	($1,278.0)	$667.1
4	$6,250.0	($62.5)	($47,250.0)	($1,312.5)	($2,913.6)	$3,273.9	($1,309.6)	$714.3
5	$6,250.0	($62.5)	($45,937.5)	($1,312.5)	($2,834.8)	$3,352.7	($1,341.1)	$761.6
6	$6,250.0	($62.5)	($44,625.0)	($1,312.5)	($2,756.0)	$3,431.5	($1,372.6)	$808.9
36	$6,250.0	($62.5)	($5,250.0)	($1,312.5)	($392.0)	$5,795.5	($2,318.2)	$2,227.3
37	$6,250.0	($62.5)	($3,937.5)	($1,312.5)	($313.2)	$5,874.3	($2,349.7)	$2,274.6
38	$6,250.0	($62.5)	($2,625.0)	($1,312.5)	($234.4)	$5,953.1	($2,381.2)	$2,321.9
39	$6,250.0	($62.5)	($1,312.5)	($1,312.5)	($155.6)	$6,031.9	($2,412.8)	$2,369.1
40	$6,250.0	($62.5)		0	($1,312.5)	($76.8)	$6,110.7	($2,444.3)	$2,416.4

Note that equipment depreciation and net receipts less disbursements are figured into the after-tax earnings of $7.5 million. These effects remain the same under either equity or bond financing.

12-20

(a) (Headings same as Problem 12-6)

0	($14,500)				($14,500)
1	$6,700	($9,672)	($2,972)	$1,165	$7,865
2	$6,700	($3,218)	$3,482	($1,365)	$5,335
3	$6,700	($1,610)	$5,091	($1,995)	$4,705
4			NPV @ 10%		$540
5					

(b) (Headings same as Problem 12-6)

0	($14,500)						
0	$10,500						($4,000)
1	$6,700	($9,667)	($3,500)	($1,260)	($4,227)	$1,658	$3,598
2	$6,700	($3,222)	($3,500)	($840)	$2,638	($1,035)	$1,325
3	$6,700	($1,611)	($3,500)	($420)	$4,669	($1,832)	$948

The two cash flow tables indicate that Mary would earn more than 10%, i* after-tax, if she financed the computers from equity. If she finances them by borrowing, she will be able to meet the cash flow requirements, but it will take a little over a year for her to recover the down payment of $4,000, and over the 3 years she will have additional funds of only $1,871. That will not go far in reducing prices of her work for her customers.

Furthermore, it appears that the analyst has underestimated the annual maintenance, property tax and insurance expenditures. A maintenance contract on three personal computers and a printer is likely cost more than $300 a year. Insurance is likely to be about 1% of first cost, or $145. Property tax is commonly 1% of first cost or more, another $145. If these estimates are more realistic than the analyst's, it does not appear that Mary will gain much advantage by investing in the computer system.

PEE Solutions Manual

CHAPTER 13

Capital Budgeting and the Choice of
a Minimum Attractive Rate of Return

General Notes

This chapter gives the instructor the opportunity to clarify some of the crude methods of analysis commonly used, and to explain in more detail how capital budgeting can make effective use of the prospective rate of return on project proposals, either mutually exclusive project alternatives or competing projects. It is also a good place to develop the concept of "resource allocation" in an overall investment program.

It also introduces students to the various concepts that underlie the selection of a minimum attractive rate of return (MARR). These concepts have frequently been badly confused in the public press, in theoretical economics papers, and by "model builders" who find it convenient to assume one can engage in unlimited borrowing. Thus any amount of money is always available for investment. As interest rates rose in the 1970's to what U.S. citizens considered to be astronomical levels, many individuals, firms and governments reduced their borrowing. Usually such interest rate increases are accompanied by high inflation which may lead some decision makers to buy on credit now in the belief that they will repay the debt from even more inflated money. When interest rates and inflation decline, borrowers can find themselves in very uncomfortable financial situations.

Every person and every company competes in the money market for investment funds and every organization has some limit to the financial resources available to it. Therefore every organization should try to allocate its resources in a way that maximizes the benefits derived. Consequently, the concepts of <u>opportunity cost</u> and of <u>capital rationing</u> both are valid as bases for establishing some minimum attractive rate of return as a means of selecting among competing proposals.

It is important for the instructor to emphasize the fact that most of the problems discussed in this book deal with choices among alternative ways of doing some one thing or of providing a defined service. The great bulk of these decisions are made at this "design" level, where a decision is made as to which of several possible alternatives for doing the same thing will be presented to top management. At the budgeting level, it will have to compete with proposals to do other things for a place in the capital budget. Designers, draftsmen, production engineers, city engineers, etc., do not and cannot know what other opportunities are available to top management. They must be given specific instructions by top management in order to have some basic criteria by which they can judge the relative attractiveness of mutually exclusive competing alternatives. The specification of a minimum attractive rate of return, either before or after income taxes, is the best way to convey one management criterion to project analysts.

PEE Solutions Manual   Chapter 13

In addition to our emphasis on the analysis of the monetary differences among alternative proposals, there is also the emphasis, albeit not as strong, on the importance of the tabulation and consideration of irreducibles in arriving at a decision. This chapter attempts to illustrate through examples and problems the fact that the selection of a minimum attractive rate of return can be influenced by an individual's appraisal of the risk involved, personal opinions regarding accuracy of estimates, and knowledge of the applicable technology. Thus, several persons reviewing the same data may arrive at several different prospective rates to use. This is really the result of applying irreducibles to mathematical analysis.

13-1

Project	Life Years	Rate of Return %	Investment	Cumulative Investment
M	10	23.01	$25,000	$25,000
O	15	13.77	25,000	50,000
L	10	12.42	25,000	75,000
N	5	10.70	50,000	125,000
P	10	10.48	25,000	150,000
Q	12	8.91	50,000	200,000
K	8	7.33	50,000	250,000

(a) 10.7%. Invest in projects M, O, L and N.
(b) This will change the order of ranking but not the projects selected. Projects L and O will be reversed. A new i* might be 9.77% (13.77% -4%).

13-2

There is no fixed solution to this problem. The economic conditions existing at the time the problem is assigned will affect the trend of the discussion. Some specific factors that might be included are:
What is the "prime rate" banks are charging their best customers?
Are interest rates high or low compared to rates over the last several years?
What effects of the current trends in interest rates on prices of bonds are discernable?
Are tax exempt municipal bonds earning more or less than their stated interest rates? Why?
Are current interest rates reflected in stock market prices? How and why?
It is interesting to have the students review the earning rates of large corporations over a period of years by having them examine several successive annual reports or by looking up the earnings rates in several successive FORTUNE magazine's annual reports on the "500" largest firms in the USA. Most students will be quite surprised at how low the actual earnings are, but they should compare them with the going interest rates of the same periods of time in order to obtain an insight into the performance of the corporations.

13-3

Mr. and Mrs. Smith must recognize that their time is valuable, worth at least $66,000 to someone else, in fact. Their net income from both their time and their investment is only $70,000. Therefore, the difference of $4,000 is all they can attribute to the effect of their investment.

(a) If their choice is between owning the franchise and working for someone else, what return can they obtain on their $36,000 in some equally secure investment? Check the current rates on savings accounts, T-bills, and Money Market Certificates. These rates will represent the opportunities available to the Smiths.

How good is the investment in the franchise? What will its salvage value be in the future? Furniture and equipment will need to be replaced about every ten years. The restaurant will have to be redecorated periodically. The real value of the franchise will depend to a great extent on the quality of the food and service and the neighborhood. It is not unreasonable to assume a zero salvage value after 10 years. Thus,

$(A/P, i\%, 10) = \$4,000/\$36,000 = 0.11111$

$i = 1.96\%$; This is not a very attractive investment because the Smiths can make any number of secure investments at a much higher rate.

Suppose a 20-year life is assumed:

$(A/P, i\%, 20) = \$4,000/\$36,000 = 0.11111$

$i = 9.2\%$; This looks more attractive.

(b) For a 10-year life, $(A/P, i\%, 10) = \$4,000/\$30,000 = 0.13333$

$i = 5.6\%$

For a 20-year life, $(A/P, i\%, 20) = 0.13333$

$i = 11.9\%$

This looks a little better, and may be competitive with current investment opportunities.

Another approach is to assume that some amount, perhaps $10,000, is the value of the franchise alone both now and in the future, and the remainder of the investment is in furniture and equipment which will have a zero salvage value in 10 years. Find i so that:

$PW = 0 = -\$30,000 + \$4,000(P/A, i\%, 10) + \$10,000(P/F, i\%, 10)$

$i = 8.93\%$

This assumption makes a big difference in the attractiveness of the investment in both parts of this problem.

PEE Solutions Manual   Chapter 13

13-4

Proposal	Life Years	Rate of Return %	Investment	Cumulative Investment
B	7	20.0	$70,000	$70,000
G	20	19.0	80,000	150,000
A	10	18.2	50,000	200,000
F	14	18.0	60,000	260,000
D	6	16.9	40,000	300,000
C	4	16.0	20,000	320,000
E	8	16.0	50,000	370,000

(b) i* = 18.2%
(c) Proposals A and F will be preferred to Proposal G. Thus, B, A, and F could be selected, using a total of only $180,000, however Proposals C and D are equally attractive as F. Thus, the selection might be B, A, and F, and C, giving an i* = 16% with the maximum life being 14 years for F.

13-5

The student should find out what the going rates are on T-Bills, Savings and Loan time deposits, bank time deposits, bank savings accounts (pass book accounts) and money market funds. He/she should evaluate the risk involved in each and determine whether or not to take that risk. (For example the money market funds are not insured by a Federal agency.) Then a plan should be made to maximize overall return. Without going into money market certificates, the plan probably can keep out about half of annual needs and invest that in a pass book account so that it is readily available. The remainder of first year needs ($3,750) should go into a 6-month money market, S&L, or bank time deposit, which ever pays a higher rate. The balance, $12,820 can be invested in 1-year and 2-year time deposits or in T-Bills (26 weeks), which ever pays a higher rate. The $10,000 minimum investment in T-Bills could pose a difficulty after 1.5 years.

Assuming a 6-month time deposit rate of 7% and T-Bill rate of 7.5% it can be shown that a student can have the necessary $7,500 for the first 2 and 1/2 years, but will fall short in the last half of the third year. Of course, higher rates may be available.

13-6

The letter should note current interest rates being charged for home mortgages, automobiles, and durable goods bought on time payment plans. It should recognize that interest rate changes vary over time and take into account the interest rate changes over the last few years. Especially, the existence of variable rate home mortgages and the history of the rates over a few years should be noted.

The letter should point out that government use of tax income for investments that return less than these going rates means that the tax payer is subsidizing projects for the government when he/she could use the money more effectively to reduce debts and interest payments--completely risk free investments.

The teacher should not be surprised to find a wide variation in the attitudes of students toward this problem. Some will consider low interest rates for government projects as being conservative. Some will try to justify different i^*'s for government projects based on their purposes. For example, they may propose low rates for social programs and high rates for defense projects. Some may consider redistribution of wealth as a justification for low i^*'s.

13-7

(a) Crude payout: Proposal L   Proposal M
    $16,000/$5,000   3.20 yrs
    $16,000/$3,500   4.57 yrs

(b) Rate of return:
    (A/P,i%,5) = $5,000/$16,000   i = 17% proposal L
    i = $3,500/$16,000 =   21.9% proposal M

Crude payout does not take the salvage value into consideration, with the result that it indicated the less desirable proposal as the one to select.

13-8

	Cash Flow	Before Tax	Cash Flow	After Tax	Difference
	L	M	L	M	M - L
0	($ 16,000)	($ 16,000)	($ 16,000)	($ 16,000)	$ 0
1	$ 5,000	$ 3,500	$ 9,400	$ 2,100	($ 7,300)
2	$ 5,000	$ 3,500	$ 3,000	$ 2,100	($ 900)
3	$ 5,000	$ 3,500	$ 3,000	$ 2,100	($ 900)
4	$ 5,000	$ 3,500	$ 3,000	$ 2,100	($ 900)
5	$ 5,000	$ 19,500	$ 3,000	$ 18,100	$ 15,100

Crude Payout after taxes   3.2 yrs.   4.4 yrs.

Rate of return after taxes   13.7%   13.1%

In this case, crude payout leads to the same recommendation.

13-9

```
Cash Flow Before Taxes
Yr              Proposal R              Proposal S
0               -$12,000                -$12,000
1               +   5,500               +   1,530
2               +   4,544               +   2,580
3               +   3,588               +   3,360
4               +   2,632               +   4,680
5               +   1,676               +   5,730
6               +     720               +   6,780

Crude Payout    2+ years                4- years

Rate of return:
R:   PW = 0 = -$12,000 +$5,500(P/A,i%,6) -$956(P/G,i%,6)
     i = 20%
S:   PW = 0 = -$12,000 +$1,530(P/A,i%,6) +$1,050(P/G,i%,6)
     i = 20%
```

In this case, the rates of return are identical. Since R has the shorter payout period, it is preferable. This conclusion could have been reached simply by observing that the cash flow is greater in the early years under Proposal R. Both conclusions assume that quick recovery of cash is desirable.

PEE Solutions Manual   Chapter 13

13-10

Proj.	Invest-ment	Annual Savings	Gross RoR	Alt.	Incremental Analysis Invest.	Savings	RoR
A1	$20,000	$4,000	15.1%				
A2	28,000	8,100	26.1%				
A3	34,000	9,300	24.2%	A3-A2	$6,000	$1,200	15.1%
B1	12,000	3,420	25.6%				
B2	16,000	4,380	24.3%	B2-B1	4,000	960	20.2%
C1	16,000	2,640	10.3%				
C2	21,000	4,620	17.7%				
D1	12,000	2,400	15.1%				
D2	16,000	4,160	22.6%				
D3	19,000	4,760	21.5%	D3-D2	3,000	600	15.1%
D4	23,000	5,980	22.6%	D4-D2	7,000	1,820	22.6%
E1	22,000	7,260	30.7%				
E2	30,000	9,300	28.5%	E2-E1	8,000	2,040	22.0%

Rankings by Rate of Return:

Alternative	Investment	RoR(%)	Total Investment
E1	$22,000	30.7	$22,000
A2	28,000	26.1	50,000
B1	12,000	25.6	62,000
D2	16,000	22.6	78,000
D4-D2	7,000	22.6	85,000
E2-E1	8,000	22.0	93,000
B2-B1	4,000	20.2	97,000
C2	21,000	17.7	118,000
A3-A2	6,000	15.1	124,000

(a) With a MARR of 20%, the budget would contain Projects A2, B2, D4, and E2 for a total investment of $97,000.

(b) With a budget constraint of $85,000, the budget would contain Projects A2, B1, D4, and E1. The implied MARR is 22.6%. If the long range prospects are that budget constraints will effectively limit investment, the company should either consider loosening its borrowing restrictions or increasing its MARR before taxes.

13-11

In order to analyze the value of the lease, it is necessary to evaluate the potential courses of action with and without the lease. With the lease, project E2-E1 may be undertaken. The cash flow differences are as follows:

Yr.	A A2 w/o lease	B A2 with lease	C Increment (E2-E1)	D w/ lease B+C	E Difference D-A
0	-$28,000	-$20,000	-$8,000	-$28,000	$ 0
1-4	+ 8,100	+ 5,100	+ 2,040	+ 7,140	-960
5-9	+ 8,100	+ 6,100	+ 2,040	+ 8,140	+ 40
10	+ 8,100	+ 8,100	+ 2,040	+ 10,140	+2,040
	+$53,000	+$39,000	+$12,400	+$51,400	-$1,600
RoR	26.1%	24.4%	22.0%	23.8%	

The lease venture will show a gross loss (PW @ 0%) of $1,600 even though $8,000 investment capital would be freed to invest in a project increment (E2-E1) promising a 22.0% return. The fact that $4,000 for the first year lease would come from operating capital does not alter the fact it is cash and must be considered as a drain on the company's capital resources.

13-12

(a) Crude Payout:

	Proposal A	Proposal B	Proposal C
$20,000/$7,000 =	2.86 yrs		
$20,000/$4,000 =		5.0 yrs	
$20,000/$2,600 =			7.69 yrs

(b) Rate of return:
(A/P,i%,4) = $7,000/$20,000
             i = 14.96%
(A/P,i%,5) = $4,000/$20,000;   i = 0%
(A/P,i%,20) = $2,600/$20,000;   i = 11.54%

In this example, the crude payout method ranks the Proposal A correctly, but reverses the ranking of the other two. This and the previous problems involving crude payout prove that, in spite of the fact that it occasionally gives the correct answer, it cannot be relied upon to give the correct answer consistently.

13-13 and 13-14

The charts are too small to make a significant difference in the estimates of the prospective rates of return. Problem, 13-13 appears to have an i of about 8% and 13-14 appears to have an i of about 9%. A larger scale chart would be a little more helpful.

The Profitability Index forms for these problems follow on the next two pages.

PEE Solutions Manual   Chapter 13

13-13   Profitability Index Form - End-of-year convention

T	0%		10%		25%		40%
-3		1.33		1.95		2.74	
-2	($ 400)	1.21	($ 484)	1.56	($ 624)	1.96	($ 784)
-1	($ 1,200)	1.1	($ 1,320)	1.25	($ 1,500)	1.4	($ 1,680)
0	($ 1,350)	1	($ 1,350)	1	($ 1,350)	1	($ 1,350)
1		0.909		0.8		0.714	
Total	($ 2,950)		($ 3,154)		($ 3,474)		($ 3,814)
			Receipts and Cost Reductions				
1	$ 150	0.909	$ 136	0.8	$ 120	0.714	$ 107
2	$ 240	0.826	$ 198	0.64	$ 154	0.51	$ 122
3	$ 330	0.751	$ 248	0.512	$ 169	0.364	$ 120
4	$ 420	0.683	$ 287	0.41	$ 172	0.26	$ 109
5	$ 420	0.621	$ 261	0.328	$ 138	0.186	$ 78
6	$ 420	0.565	$ 237	0.262	$ 110	0.133	$ 56
7	$ 420	0.513	$ 215	0.21	$ 88	0.095	$ 40
8	$ 420	0.467	$ 196	0.168	$ 71	0.068	$ 29
9	$ 420	0.424	$ 178	0.134	$ 56	0.048	$ 20
10	$ 420	0.386	$ 162	0.107	$ 45	0.035	$ 15
11	$ 420	0.351	$ 147	0.086	$ 36	0.025	$ 11
12	$ 420	0.319	$ 134	0.069	$ 29	0.018	$ 8
13	$ 420	0.29	$ 122	0.055	$ 23	0.013	$ 5
14	$ 300	0.263	$ 79	0.044	$ 13	0.009	$ 3
15	$ 880	0.239	$ 210	0.035	$ 31	0.006	$ 5
Total	$ 6,100		$ 2,810		$ 1,255		$ 728
Ratio	0.4836		1.1224		2.7681		5.2390

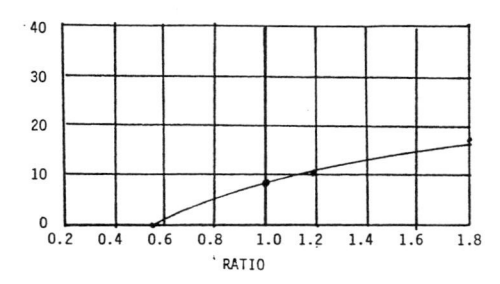

13-14 Profitability Index Form - End-of-year convention

T	0%		10%		25%		40%
-3		1.33		1.95		2.74	
-2	($200)	1.21	($242)	1.56	($312)	1.96	($392)
-1	($1,200)	1.1	($1,320)	1.25	($1,500)	1.4	($1,680)
0	($1,350)	1	($1,350)	1	($1,350)	1	($1,350)
1		0.909		0.8		0.714	
Total	($2,750)		($2,912)		($3,162)		($3,422)
		Receipts and Cost Reductions					
1	$150	0.909	$136	0.8	$120	0.714	$107
2	$240	0.826	$198	0.64	$154	0.51	$122
3	$330	0.751	$248	0.512	$169	0.364	$120
4	$420	0.683	$287	0.41	$172	0.26	$109
5	$420	0.621	$261	0.328	$138	0.186	$78
6	$420	0.565	$237	0.262	$110	0.133	$56
7	$420	0.513	$215	0.21	$88	0.095	$40
8	$420	0.467	$196	0.168	$71	0.068	$29
9	$420	0.424	$178	0.134	$56	0.048	$20
10	$420	0.386	$162	0.107	$45	0.035	$15
11	$420	0.351	$147	0.086	$36	0.025	$11
12	$420	0.319	$134	0.069	$29	0.018	$8
13	$420	0.29	$122	0.055	$23	0.013	$5
14	$300	0.263	$79	0.044	$13	0.009	$3
15	$812	0.239	$194	0.035	$28	0.006	$5
Total	$6,032		$2,794		$1,252		$728
Ratio	0.4559		1.0422		2.5256		4.7005

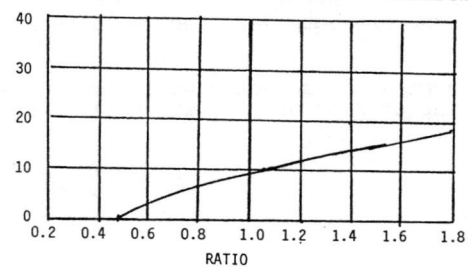

PEE Solutions Manual   Chapter 13

13-15
Capital available = $250,000; $i^* = 18\%$ before income tax

Project	Investment	Annual Savings	RoR	Incremental RoR	Accept or reject
A1	$50,000	$12,000	20.2%		Accept
A2	100,000	25,000	21.4		Accept
A3	125,000	27,000	17.2		Reject
B1	60,000	14,800	21.0		Accept
B2	80,000	16,500	15.9		Reject
C1	40,000	8,000	15.1		Reject
C2	50,000	12,000	20.2		Accept
C3	75,000	16,000	16.8	9.6% on C3-C2	Reject
C4	85,000	20,000	19.6	18.8% on C4-C2	Accept
D1	30,000	7,000	19.4		Accept
D2	38,000	8,000	16.5		Reject
D3	47,000	11,000	19.4	19.4% on D3-D1	Accept
E1	43,000	10,000	19.3		Accept
E2	60,000	14,500	20.4		Accept
E3	70,000	15,700	18.2	3.5% on E3-E2	Reject

(a)
Proj.	Investment	RoR	Cumulative Investment
A2	$100,000	21.4%	$100,000
B1	60,000	21.0	160,000
E2	60,000	20.4	220,000
C2	50,000	20.2%	270,000
D1	30,000	19.4	300,000
D3-D1	17,000	19.4	317,000
C4-C2	35,000	18.8	352,000

(b) Without borrowing, the projects selected would be A2, B1, E2, and D1, for a total of $250,000.

(c) With borrowing, the projects selected would be A2, B1, E2, D3, and C4 for a total investment of $352,000, of which $102,000 would be borrowed at 10%.

PEE Solutions Manual

Chapter 13

13-16 Profitability Index Form - Continuous Compounding

T	0%	F/P	10%	F/P	25%	F/P	40%
-3		1.29		1.74		2.3	
-2		1.15		1.39		1.65	
-1		1.05		1.12		1.18	
0	($80,000)	1	($80,000)	1	($80,000)	1	($80000)
1		0.954		0.896		0.849	
Total	($80,000)		($80,000)		($80,000)		($80000)
		P/F		P/F		P/F	
1	$22,000	0.954	$20,988	0.896	$19,712	0.849	$18,678
2	$23,800	0.867	$20,635	.717	$17,065	0.607	$14,447
3	$25,600	0.788	$20,173	0.574	$14,694	0.433	$11,085
4	$25,800	0.717	$18,499	0.459	$11,842	0.31	$7,998
5	$26,000	0.652	$16,952	0.367	$9,542	0.221	$5,746
6	$25,000	0.592	$14,800	0.294	$7,350	0.158	$3,950
7	$20,000	0.538	$10,760	0.235	$4,700	0.113	$2,260
8	$36,000	0.49	$17,640	0.188	$6,768	0.081	$2,916
9		0.445	$0	0.15	$0	0.058	$0
Total	$204,200		$140,446		$91,673		$67,079
Ratio	0.3918		0.5696		0.8727		1.1926

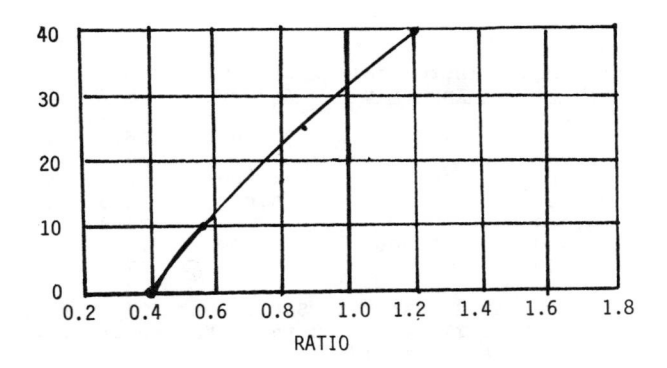

13-17

```
Common Stock       $45,000,000 at 13.33% after-tax or 22.2% before-tax
Short term notes   $2,000,000  at 14.0%
Long term debt     $15,000,000 at 10.0%
Total Capital  =   $62,000,000
Cost of Capital:         45/62 @ 22.2% = 0.1611
                          2/62 @ 14.0% = 0.0045
                         15/62 @ 10.0% = 0.0242
                Cost of Capital before-tax = 0.1898 or 18.98%
```

The drop in share prices from $58 to $45 indicates either or both of two things: (1) the earnings prospects for the company have fallen, (2) interest rates are rising. The fact that general stock market price levels have been dropping slightly indicates that earnings rates needed to attract equity capital are rising. Therefore, the cost of capital in the future is likely to be higher, and the i* to be used in project proposals at this time ought to be raised. Since the cost of capital is likely to rise, the before-tax i* should be in the order of 22% to 25%.

13-18

To sell 119,048 shares at $42 to raise the $5,000,000 needed, the company would dilute the value of all the original one million shares, and suffer a market drop of $3,000,000. If the project is realistically analyzed and shows a prospective rate of return of 13% after income taxes, the company should have no trouble in meeting the cash flow problems of repaying the $5 million 5-year loan with an after-tax cost of capital of about 7.8%. That should enhance the earnings on equity and result in higher share prices in the market place. It appears that this is a good example of "borrowing making good business better." The 10-year bond might be rejected because the repayment goes beyond the effects of the investment by 4 years, and results in payment of more total interest than would be incurred on the 5-year loan. However, "separable decisions should be made separately."

13-19

Prop. No.	Total Invest't	Net Annual Cash Flow	Increm'tal Invest't	Increm'tal Income	Overall RoR	RoR on Increment	Life Years
A1	$25,000	$6,200			18.4%		8
A2	35,000	8,200	$10,000	$2,000	15.8	11.8% Rej	8
B1	40,000	12,000			19.9		6
B2	60,000	16,500	20,000	4,500	16.5	9.31% Rej	6
B3	85,000	24,500	25,000	8,000	18.3	16.9% Ac	6
C1	35,000	9,300			23.3		10
C2	45,000	10,300	10,000	1,000	18.8	0.0% Rej	10
D1	50,000	14,000			17.2		6

Select C-1, $35,000 at 23.3%; B-1, $40,000 at 19.9%; A-1, $25,000 at 18.4%; and D-1, $50,000 at 17.2%. That takes the entire capital budget of $150,000 with an implied i* of 17.2%.

PEE Solutions Manual   Chapter 13

13-20
   In this case C-1 is more than 2% higher than B-1 and more than 1% higher than A-1 and D-1. Also, A-1 is more than 1% higher than D-1. There is no better combination than that already selected. For example, B-3, at 18.3% overall, might be chosen over D-1, but that would involve investing $45,000 at an incremental rate of only 16.9%, less than the return offered by D-1. The final choice will depend upon how strongly management believes that the 1%-for-2years rule should be applied.

13-21
   Yes. The incremental rate of return on the additional investment of Proposal B-3 over B-1 will earn 16.9%. That seems adequate to justify borrowing $45,000 at 12% (7.2% after taxes) in order to finance it. Proposals A-2 and C-2 will not earn enough on their incremental investments to justify borrowing. Therefore the only change in the selection of Problem 13-19 would be substitute B-3 for B-1.

13-22
   Proposal A: n = 6; P = -$18,000; A = $5,000; G = -$300; S = 0
   Payout by successive years: $5,000 +$4,700 +$4,400 +$4,100 = $18,200. Thus, crude payout is achieved in a little less than 4 years.
      Find i so that NPW = 0 = -$18,000 +$5,000(P/A,i%,6)
         -$300(P/G,i%,6) ; By interpolation, i = 11.75%

   Proposal B: n = 8; P = -$18,000; A = $3,500; S = +$4,000
   Payout = $18,000/$3,500 = 5.14 years
      Find i so that NPW = 0 = -$18,000 +$3,500(P/A,i%,8)
         +$4,000(P/F,i%,8);   i = 13.45%

   Neither proposal meets both criteria. Clearly Proposal B is more attractive than A, but it may be rejected if the payout criterion is deemed to be more significant than the rate of return. Any significant salvage value is ignored in the payout method but may be very important in the rate of return method of analysis.

13-23
   No solution can be given for this suggested term report.

155

PEE Solutions Manual

CHAPTER 14

Prospective Inflation and Sensitivity Analysis

14-1
NPW = 0 = -$9,600 + $650(P/A,i%,10) + $10,000(P/F,i%,10)
from which i = 7.07%.
The following analysis, along the lines of Table 14-2, includes the effect of inflation and of the $100 capital gains tax in 1987.

Year	Cash Flow in current dollars	Consumer Price Index	Cash flow in 1977 dollars
1977	($9,600)	60.6	($9,600)
1978	650	65.2	604
1979	650	72.6	543
1980	650	82.5	477
1981	650	90.9	433
1982	650	96.5	408
1983	650	99.6	395
1984	650	103.9	379
1985	650	107.6	366
1986	650	109.6	359
1987	10,550	113.6	5,628
Total	$6,800		($6)
Rate of Return	7.00%		-0.01%

PEE Solutions Manual   Chapter 14

14-2
Annual interest payment received after income tax =
0.14($10,000)(1 -0.28) = $1,008

Year	Cash Flow Current	Cash Flow Constant	Year	Cash Flow Current	Cash Flow Constant
0	($10,000)	($10,000)	0	($10,000)	($10,000)
1	1,008	960	1	1,008	960
2	1,008	914	2	1,008	914
3	1,008	871	3	1,008	871
4	1,008	829	4	1,008	829
5	1,008	790	5	1,008	790
6	1,008	752	6	1,008	752
7	1,008	716	7	1,008	716
8	1,008	682	8	1,008	682
9	1,008	650	9	1,008	650
10	11,008	6,758	10	1,008	619
Sums	$10,080	$3,923	11	1,008	589
			12	1,008	561
			13	1,008	535
			14	1,008	509
			15	1,008	485
			16	1,008	462
			17	1,008	440
			18	1,008	419
			19	1,008	399
			20	11,008	4,149
			Sums	$20,160	$6,331
RoR	10.1%		4.8%	10.1%	4.8%

14-3

Year	Taxable Income	Tax	Tax percentage	Tax rate on increment of investment
0	$45,000	$6,750	15.00%	15%
1	51,750	7,938	15.34	25%
2	59,512	9,878	16.60	25%
3	68,439	12,110	17.69	25%
4	78,705	15,010	19.07	34%
5	90,511	19,024	21.02	34%
6	104,088	23,844	22.91	39%

PEE Solutions Manual   Chapter 14

14-4
     Under the stated conditions a 50% rise in the price level would lead
to a taxable income of $60,000 with an incremental tax rate of 25%. A
rise of 100% would lead to a taxable income of $80,000 with an
incremental tax rate of 34%. A rise of 150% would lead to $100,000.
This final increment of income also would be taxed at 34%.

14-5
     (a) The $1,000 a year positive gradient in cash flow due to the
reduction in labor costs will be partially offset by a negative gradient
of $400 a year caused by increased income taxes.
     The equation to find before-tax rate of return becomes:
     0 = -$110,000 +$26,300(P/A,i%,10) +$1,000(P/G,i%,10)
     i = approximately 23.8% before income taxes
     The equation to find after-tax rate of return becomes:
     0 = -$110,000 +$20,180(P/A,i%,10) +($1,000 -$400)(P/G,i%,10)
     i = approximately 15.3%
     The prospect of the favorable differential price change increases the
prospective before-tax rate of return by 3.7% (i.e., 23.8% - 20.1%) and
increases the prosepctive after-tax rate by 2.4% (i.e., 15.3% - 12.9%).
     (b) The answer to this question depends on whether the prospective
before-tax gradient of $1,000 a year is an expected <u>differential</u> price
change or whether it is expected to take place in part or entirely
because of general inflation. The answer in (a) would not be changed if
a differential price change is expected. If the gradient should be
anticipated in part or entirely because of general inflation, the
differences found in (a) would be reduced or eliminated.

14-6
     For reasons brought out in Example 14-3, during periods of inflation
the prospective rate of return on an investment will be reduced where
depreciation deductions to determine taxable income are to be based on
the investment at zero date with no adjustment for changes in the
purchasing power of the monetary unit. It follows that if an economy
study is to be made in monetary units of constant purchasing power and if
a value of i* has been stipulated, a recognition of this rule of income
taxation will cause some increments of investment to be rejected that
would be acceptable if the rule were not recognized. As applied to this
choice among competing wire sizes, the forecast of inflation with no
differential price change in the cost of energy would tend to favor lower
investments and therefore <u>might</u> lead to selection of a smaller wire size
than would otherwise be chosen.

PEE Solutions Manual   Chapter 14

14-7

(a) $0 = -\$100,000 + \$300,000(P/F,i\%,10)$

i = approximately 11.6%, the apparent rate of return before income taxes if the effect of inflation is not recognized.

(b) On date 10 the tax on the so-called profit is
$0.28(\$300,000 - \$100,000) = \$56,000$. The net cash flow after this tax is $\$300,000 - \$56,000 = \$244,000$

$0 = -\$100,000 + \$244,000(P/F,i\%,10)$

i = approximately 9.3%, the apparent rate of return after income taxes if the effect of inflation is neglected.

(c) Because the price level has doubled, the cash flow at date 10 must be multiplied by 0.5 to convert it to zero-date dollars.

$0 = -\$100,000 + \$244,000(0.5)(P/F,i\%,10)$

i = approximately 2.0% after taxes if the effect of inflation is recognized.

14-8

Of course the answer to (a) is unchanged.

(b) The tax is $0.15(\$300,000 - \$100,000) = \$30,000$

$0 = -\$100,000 + \$270,000(P/F,i\%,10)$

i = approximately 10.5%

(c) $0 = -\$100,000 + \$270,000(0.5)(P/F,i\%,10)$

i = approximately 3.0%

14-9

In 14-7(c) the tax will be $0.28(\$100,000) = \$28,000$

$0 = -\$100,000 + \$272,000(0.5)(P/F,i\%,10)$

i = approximately 3.1%

In 14-8(c) the tax will be $0.15(\$100,000) = \$15,000$

$0 = -\$100,000 + \$285,000(0.5)(P/F,i\%,10)$

i = approximately 3.6%

PEE Solutions Manual   Chapter 14

14-10
See spreadsheets for 14-11 and 14-12.

14-11

ANALYSIS OF A BOND INVESTMENT AFTER INCOME TAXES

Purchase price of bond $10,000 Coupon interest rate
Years to maturity   10 Inflation rate
Investor's incremental tax rate   30.00%

Year	Before-tax cash flow in current dollars	Cash flow for income taxes	After-tax cash flow in current dollars	Estimated price index assuming 100 for zero date	After-tax cash flow in zero-date dollars
0	-10000		-10000	100.00	-10000
1	1200	-360	840	108.00	778
2	1200	-360	840	116.64	720
3	1200	-360	840	125.97	667
4	1200	-360	840	136.05	617
5	1200	-360	840	146.93	572
6	1200	-360	840	158.69	529
7	1200	-360	840	171.38	490
8	1200	-360	840	185.09	454
9	1200	-360	840	199.90	420
10	11200	-360	10840	215.89	5021
Total	12000	-3600	8400		268
Rate of return	12.00%		8.40%		0.37%

PEE Solutions Manual   Chapter 14

14-12

ANALYSIS OF A BOND INVESTMENT AFTER INCOME TAXES

Purchase price of bond $10,000 Coupon interest rate
Years to maturity   10 Inflation rate
Investor's incremental tax rate   30.00%

Year	Before-tax cash flow in current dollars	Cash flow for income taxes	After-tax cash flow in current dollars	Estimated price index assuming 100 for zero date	After-tax cash flow in zero-date dollars
0	-10000		-10000	100.00	-10000
1	1200	-360	840	104.00	808
2	1200	-360	840	108.16	777
3	1200	-360	840	112.49	747
4	1200	-360	840	116.99	718
5	1200	-360	840	121.67	690
6	1200	-360	840	126.53	664
7	1200	-360	840	131.59	638
8	1200	-360	840	136.86	614
9	1200	-360	840	142.33	590
10	11200	-360	10840	148.02	7323
Total	12000	-3600	8400		3569
Rate of return	12.00%		8.40%		4.23%

14-13

See spreadsheets for Problems 14-14 and 14-15.

161

PEE Solutions Manual   Chapter 14

14-14

Initial investment $110,000 Est life (years)   25
Inflation rate   12.0%

Year	Cash flow before income taxes in zero-date dollars A	Estimated price index assuming 100 for zero date B	Estimated before-tax cash flow in current dollars BA/100 C	Write-off of initial outlay for tax purposes D	Taxable income from project (C + D) E	Cash flow for income taxes in current dollars -0.40E F	Cash flow after income taxes in current dollars (C + F) G	Cash flow after income taxes in zero-date dollars 100G/B H
0	-110000	100.00	-110000				-110000	-110000
1	26300	112.00	29456	-4400	25056	-10022	19434	17351
2	26300	125.44	32991	-4400	28591	-11436	21554	17183
3	26300	140.49	36950	-4400	32550	-13020	23930	17033
4	26300	157.35	41384	-4400	36984	-14793	26590	16899
5	26300	176.23	46350	-4400	41950	-16780	29570	16779
6	26300	197.38	51912	-4400	47512	-19005	32907	16672
7	26300	221.07	58141	-4400	53741	-21496	36645	16576
8	26300	247.60	65118	-4400	60718	-24287	40831	16491
9	26300	277.31	72932	-4400	68532	-27413	45519	16415
10	26300	310.58	81684	-4400	77284	-30914	50770	16347
11		347.85		-4400	-4400	1760	1760	506
12		389.60		-4400	-4400	1760	1760	452
13		436.35		-4400	-4400	1760	1760	403
14		488.71		-4400	-4400	1760	1760	360
15		547.36		-4400	-4400	1760	1760	322
16		613.04		-4400	-4400	1760	1760	287
17		686.60		-4400	-4400	1760	1760	256
18		769.00		-4400	-4400	1760	1760	229
19		861.28		-4400	-4400	1760	1760	204
20		964.63		-4400	-4400	1760	1760	182
21		1080.38		-4400	-4400	1760	1760	163
22		1210.03		-4400	-4400	1760	1760	145
23		1355.23		-4400	-4400	1760	1760	130
24		1517.86		-4400	-4400	1760	1760	116
25		1700.01		-4400	-4400	1760	1760	104
Total	153000		406916	-110000	406916	-162766	244149	61604

Rate of
Return   20.07% before income taxes   after income taxes   8.83%

162

PEE Solutions Manual   Chapter 14

14-15

Initial investment $110,000 Est life (years)   10
Inflation rate   6.0%

Year	Cash flow before income taxes in zero-date dollars A	Estimated price index assuming 100 for zero date B	Estimated before-tax cash flow in current dollars BA/100 C	Write-off of initial outlay for tax purposes D	Taxable income from project (C + D) E	Cash flow for income taxes in current dollars -0.40E F	Cash flow after income taxes in current dollars (C + F) G	Cash flow after income taxes in zero-date dollars 100G/B H
0	-110000	100.00	-110000				-110000	-110000
1	26300	106.00	27878	-31460	-3582	1433	29311	27652
2	26300	112.36	29551	-22440	7111	-2844	26706	23769
3	26300	119.10	31324	-16060	15264	-6105	25218	21174
4	26300	126.25	33203	-11440	21763	-8705	24498	19405
5	26300	133.82	35195	-9570	25625	-10250	24945	18641
6	26300	141.85	37307	-9570	27737	-11095	26212	18479
7	26300	150.36	39545	-9460	30085	-12034	27511	18297
8	26300	159.38	41918		41918	-16767	25151	15780
9	26300	168.95	44433		44433	-17773	26660	15780
10	26300	179.08	47099		47099	-18840	28260	15780
Total	153000		257454.2014	-110000	257454	-102982	154473	84754

Rate of
Return   20.07% before income taxes   after income taxes   13.58%

14-16

Initial investment $110,000 Est life (years) 10
Inflation rate 12.0%

Year	Cash flow before income taxes in zero-date dollars A	Estimated price index assuming 100 for zero date B	Estimated before-tax cash flow in current dollars BA/100 C	Write-off of initial outlay for tax purposes D	Taxable income from project (C + D) E	Cash flow for income taxes in current dollars -0.40E F	Cash flow after income taxes in current dollars (C + F) G	Cash flow after income taxes in zero-date dollars 100G/B H
0	-110000	100.00	-110000				-110000	-110000
1	26300	112.00	29456	-22000	7456	-2982	26474	23637
2	26300	125.44	32991	-22000	10991	-4396	28594	22795
3	26300	140.49	36950	-22000	14950	-5980	30970	22044
4	26300	157.35	41384	-22000	19384	-7753	33630	21373
5	26300	176.23	46350	-22000	24350	-9740	36610	20773
6	26300	197.38	51912		51912	-20765	31147	15780
7	26300	221.07	58141		58141	-23256	34885	15780
8	26300	247.60	65118		65118	-26047	39071	15780
9	26300	277.31	72932		72932	-29173	43759	15780
10	26300	310.58	81684		81684	-32674	49010	15780
Total	153000		406915.5402	-110000	406916	-162766	244149	79522

Rate of Return   20.07% before income taxes   after income taxes   12.64%

14-17

Let n be the number of years of life to break even for the new material.

$3,000(A/P,12\%,3) = $5,800(A/P,12\%,n)$

$(A/P,12\%,n) = \dfrac{\$3,000}{\$5,800}(0.41635) = 0.21535$

Interpolation between the capital recovery factors for 7 and 8 years (0.21912 and 0.20130, respectively) indicates that the new type of lining must last approximately 7.2 years to break even on annual cost.

PEE Solutions Manual   Chapter 14

14-18
   The choice is between an extra $160,000 (i.e., $1,435,000 - $1,275,000) now and an extra $250,000 (i.e., $960,000 - $710,000) after n years. For the two to break even:
   $160,000(F/P,8%,n) = $250,000
   (F/P,7%,n) = 1.5625

   Interpolation between the 8% compound amount factors for 6 and 7 years indicates that the new story ought to be needed within approximately 6.6 years for costs to break even at an i of 8%. In this problem the 40-year life is irrelevant because it does not change any cash flow affected by the choice between the alternatives.

14-19
   See the spreadsheets for Problems 14-20, 21 and 22.

PEE Solutions Manual   Chapter 14

14-20
Parts (a) and (b)

Initial investment $110,000 Est life (years)   10
Inflation rate   12.0%

Year	Cash flow before income taxes in zero-date dollars A	Estimated price index assuming 100 for zero date B	Estimated before-tax cash flow in current dollars BA/100 C	Write-off of initial outlay for tax purposes D	Taxable income from project (C + D) E	Cash flow for income taxes in current dollars -0.40E F	Cash flow after income taxes in current dollars (C + F) G	Cash flow after income taxes in zero-date dollars 100G/B H
0	-110000	100.00	-110000				-110000	-110000
1	22000	112.00	24640	0	24640	-9856	14784	13200
2	22000	125.44	27597	0	27597	-11039	16558	13200
3	22000	140.49	30908	0	30908	-12363	18545	13200
4	22000	157.35	34617	0	34617	-13847	20770	13200
5	22000	176.23	38772	0	38772	-15509	23263	13200
6	22000	197.38	43424	0	43424	-17370	26054	13200
7	22000	221.07	48635	0	48635	-19454	29181	13200
8	22000	247.60	54471	0	54471	-21788	32683	13200
9	22000	277.31	61008	0	61008	-24403	36605	13200
10	132000	310.58	409972	0	299972	-27331	289983	93367
10	110000				231643	-92657		
Total	220000		664044	0	664044	-265618	398426	102167

                        Before Income Taxes   After Income Taxes
Rate of   Constant $ Current $ Current $ Constant $
Return   20.00%   34.40%   23.56%   10.32%

(c) If the spreadsheet is set up efficiently, it will only be necessary to change one figure. The lower portion is shown below.

6	22000	197.38	43424	0	43424	-17370	26054	13200
7	22000	221.07	48635	0	48635	-19454	29181	13200
8	22000	247.60	54471	0	54471	-21788	32683	13200
9	22000	277.31	61008	0	61008	-24403	36605	13200
10	132000	310.58	409972	0	299972	-27331	336312	108283
10	110000		341643		231643	-46329		
Total	220000		664044	0	664044	-219289	444755	117083

                        Before Income Taxes   After Income Taxes
Rate of   Constant $ Current $ Current $ Constant $
Return   20.00%   34.40%   24.54%   11.20%

14-21

Parts (a) and (b) [Note: This spreadsheet is an easy modification of that from Problem 14-20.]

Initial investment $110,000   Est life (years)   10
Inflation rate   12.0%

Year	Cash flow before income taxes in zero-date dollars A	Estimated price index assuming 100 for zero date B	Estimated before-tax cash flow in current dollars BA/100 C	Write-off of initial outlay for tax purposes D	Taxable income from project (C + D) E	Cash flow for income taxes in current dollars -0.40E F	Cash flow after income taxes in current dollars (C + F) G	Cash flow after income taxes in zero-date dollars 100G/B H
0	-110000	100.00	-110000				-110000	-110000
1	22000	112.00	24640	0	24640	-9856	14784	13200
2	22000	125.44	27597	0	27597	-11039	16558	13200
3	22000	140.49	30908	0	30908	-12363	18545	13200
4	22000	157.35	34617	0	34617	-13847	20770	13200
5	22000	176.23	38772	0	38772	-15509	23263	13200
6	22000	197.38	43424	0	43424	-17370	26054	13200
7	22000	221.07	48635	0	48635	-19454	29181	13200
8	22000	247.60	54471	0	54471	-21788	32683	13200
9	22000	277.31	61008	0	61008	-24403	36605	13200
10	200698	310.58	623338	0	513338	-27331	418003	134586
10	178698		555009		445009	-178004		
Total	288698		877410	0	877410	-350964	526446	143386

Before Income Taxes   After Income Taxes
Rate of Constant $ Current $ Current $ Constant $
Return   22.16%   36.82%   26.08%   12.57%

(c) Only the lower (changed) portion of the spreadsheet is shown here.

6	22000	197.38	43424	0	43424	-17370	26054	13200
7	22000	221.07	48635	0	48635	-19454	29181	13200
8	22000	247.60	54471	0	54471	-21788	32683	13200
9	22000	277.31	61008	0	61008	-24403	36605	13200
10	200698	310.58	623338	0	513338	-27331	507004	163242
10	178698		555009		445009	-89002		
Total	288698		877410	0	877410	-261962	615448	172042

Before Income Taxes   After Income Taxes
Rate of Constant $ Current $ Current $ Constant $
Return   22.16%   36.82%   27.56%   13.89%

PEE Solutions Manual   Chapter 14

14-22
Parts (a) and (b) [Note: This spreadsheet is an easy modification of that from Problem 14-20.]

```
Initial investment   $110,000  Est life (years)        10
Inflation rate                  12.0%
```

	Cash flow before income taxes in zero-date dollars	Estimated price index assuming 100 for zero date	Estimated before-tax cash flow in current dollars	Write-off of initial outlay for tax purposes	Taxable income from project	Cash flow for income taxes in current dollars	Cash flow after income taxes in current dollars	Cash flow after income taxes in zero-date dollars
Year			BA/100		(C + D)	-0.40E	(C + F)	100G/B
	A	B	C	D	E	F	G	H
0	-110000	100.00	-110000				-110000	-110000
1	22000	112.00	24640	0	24640	-9856	14784	13200
2	22000	125.44	27597	0	27597	-11039	16558	13200
3	22000	140.49	30908	0	30908	-12363	18545	13200
4	22000	157.35	34617	0	34617	-13847	20770	13200
5	22000	176.23	38772	0	38772	-15509	23263	13200
6	22000	197.38	43424	0	43424	-17370	26054	13200
7	22000	221.07	48635	0	48635	-19454	29181	13200
8	22000	247.60	54471	0	54471	-21788	32683	13200
9	22000	277.31	61008	0	61008	-24403	36605	13200
10	92835	310.58	288331	0	178331	-27331	216999	69868
10	70835		220003		110003	-44001		
Total	180835		542404	0	542404	-216961	325442	78668
		Before Income Taxes					After Income Taxes	
Rate of	Constant $		Current $				Current $	Constant $
Return	18.52%		32.75%				21.83%	8.77%

(c) Only the lower (changed) portion of the spreadsheet is shown here.

6	22000	197.38	43424	0	43424	-17370	26054	13200
7	22000	221.07	48635	0	48635	-19454	29181	13200
8	22000	247.60	54471	0	54471	-21788	32683	13200
9	22000	277.31	61008	0	61008	-24403	36605	13200
10	92835	310.58	288331	0	178331	-27331	238999	76951
10	70835		220003		110003	-22001		
Total	180835		542404	0	542404	-194961	347443	85751
		Before Income Taxes					After Income Taxes	
Rate of	Constant $		Current $				Current $	Constant $
Return	18.52%		32.75%				22.38%	9.27%

PEE Solutions Manual   Chapter 14

14-20 to 14-22 comments
These three problems concentrate attention on what happens when taxable capital gains are calculated without recognition of changes in the value of the monetary unit. The adverse effect of this on investors during a period of substantial inflation is apparent. Also it is evident that there is an equitable basis for taxing capital gains computed in this usual way at a lower rate (20% in these problems) than the rate applied to ordinary income (40% in these problems.)

The following matrix shows the relevant figures for calculated rates of return

	Conventional analysis –no inflation considered	Analysis in terms of zero-date dollars with 12% annual inflation and differential price change only in resale price of land		
		No d.p.c.	Favorable d.p.c.	Unfavorable d.p.c.
Before taxes	20.0%	20.0%	22.2%	18.5%
After taxes with 40% tax rate on all taxable income	12.0%	10.3%	12.6%	8.8%
After taxes with 40% tax rate on ordinary income and 20% rate on capital gains	12.0%	11.2%	13.9%	9.3%

It should be noted that problems 14-20, 14-21, and 14-22 do not involve the distortions introduced into the calculation of taxable income that are caused by basing depreciation deductions on costs at a different price level. Because Example 9-1 did not involve any assets depreciable for tax purposes, depreciation does not enter into these three problems. Moreover, because it was assumed that all other relevant cash flows would increase at the same rate as general price levels, the only differential price changes are those from the sale of the land at the end of the 10 years and from the income taxes based on that sale.

14-23

```
Initial investment    $15,000  Est life (years)        10
Inflation rate        6.0%
```

	Cash flow before income taxes in zero-date dollars	Estimated price index assuming 100 for zero date	Estimated before-tax cash flow in current dollars	Write-off of initial outlay for tax purposes	Taxable income from project	Cash flow for income taxes in current dollars	Cash flow after income taxes in current dollars	Cash flow after income taxes in zero-date dollars
Year			BA/100		(C + D)	-0.40E	(C + F)	100G/B
	A	B	C	D	E	F	G	H
0	-15000	100.00	-15000				-15000	-15000
1	4100	106.00	4346	-1500	2846	-1138	3208	3026
2	4100	112.36	4607	-1500	3107	-1243	3364	2994
3	4100	119.10	4883	-1500	3383	-1353	3530	2964
4	4100	126.25	5176	-1500	3676	-1470	3706	2935
5	4100	133.82	5487	-1500	3987	-1595	3892	2908
6	4100	141.85	5816	-1500	4316	-1726	4090	2883
7	4100	150.36	6165	-1500	4665	-1866	4299	2859
8	4100	159.38	6535	-1500	5035	-2014	4521	2836
9	4100	168.95	6927	-1500	5427	-2171	4756	2815
10	4100	179.08	7342	-1500	5842	-2337	5005	2795
Total	26000		42284	-15000	42284	-16913	25370	14016

```
                 Before Income Taxes              After Income Taxes
Rate of   Constant $        Current $         Current $       Constant $
Return     24.21%            31.66%            21.35%          14.48%
```

The rate of return is reduced from 15.6% to 14.5% as the result of the "modest" 6% inflation rate.

PEE Solutions Manual										Chapter 14

14-24
Rates of return under the various inflation rate assumptions are:

	General (High) Labor			General (Low) Labor	
Energy	High	Low	Energy	High	Low
H	8.0	4.8	H	12.3	<u>8.9</u>
L	9.9	6.9	L	14.4	<u>11.1</u>

If the general inflation rate is high, there is no case under which the project is acceptable. The underlined rates are those with any reasonable likelihood of occurring. Thus, of the cases studied, it is only under the condition of low general labor and energy inflation rates that the project is acceptable. A sample of the spreadsheet follows.

Year	Cash Flows in Year 0 Dollars				Cash flow before income taxes	Write-off of initial outlay for tax purposes	Cash flow for income taxes	Cash flow after income taxes
	Labor savings	Material savings	Energy disburse.	Maint. disburse.				
1	17585	5000	-3263	-1200	18123	-9217	-3562	14560
2	17180	5000	-3548	-1200	17432	-8495	-3575	13857
3	16784	5000	-3859	-1200	16725	-7829	-3558	13167
4	16397	5000	-4197	-1200	16001	-7216	-3514	12487
5	16020	5000	-4564	-1200	15255	-6650	-3442	11813
6	15651	5000	-4964	-1200	14487	-6129	-3343	11144
Totals	99617	30000	-24395	-7200	98022	-45536	-20994	77027

PEE Solutions Manual　　　　　　　　　　　　　　　　　　　　　　　　　　　　　Chapter 14

14-24(b)

```
Initial Investment  $60,000  Est. Life     6 years
Annual Savings:  Labor:     $18,000  Material:  $5,000
An. Disbursements:Energy:   $3,000   Maint.:    $1,200
Income Tax Rate:     40%
Inflation Rates: General:      8.5%  Labor:      6.0%
                 Energy:      18.0%
```

Development of Cash Flows in Current Dollar Amounts

Year	Labor	Material	Energy	Maint'ce	Total
Base 0	18000	5000	-3000	-1200	18800
1	19080	5425	-3540	-1302	19663
2	20225	5886	-4177	-1413	20521
3	21438	6386	-4929	-1533	21363
4	22725	6929	-5816	-1663	22175
5	24088	7518	-6863	-1804	22939
6	25533	8157	-8099	-1958	23634

Year	Cash flow before income taxes (current dollars) A	Write-off of initial outlay for tax purposes B	Influence on taxable income (A - B) C	Cash flow for income taxes -0.40C D	Cash flow after income taxes (current dollars) (A + D) E	Cash flow after income taxes (constant dollars) F
0	-60000				-60000	-60000
1	19663	-10000	9663	-3865	15798	14560
2	20521	-10000	10521	-4208	16313	13857
3	21363	-10000	11363	-4545	16818	13167
4	22175	-10000	12175	-4870	17305	12487
5	22939	-10000	12939	-5175	17763	11813
6	23634	-10000	13634	-5454	18181	11144
Totals	70294	-60000	70294	-28118	42177	17027
Rate of Return	26.8%				17.2%	8.0%

172

PEE Solutions Manual   Chapter 14

14-25

	2,500 Gal/Day	5,000 Gal/Day
PW of costs for 15 years		
First cost	#32,000	#48,000

#32,000(P/F,15%,5) =   15,846
Less PW of residual value
   #32,000(A/P,15%,15)(P/A,15%,5)(P/F,15%,15) = -2,255

Operation, Maintenance, & Repairs:
   #1,000(P/A,15%,15) -#500(P/A,15%,5) =   4,171
   #750(P/A,15%,15) =   4,385

Electric Energy:
   #20[1,000(P/A,15%,15) +#350(P/G,15%,11)
   +#3,500(P/A,15%,4)(P/F,15%,11)]365/100,000 =   1,072
   #15(14,690)365/100,000 =   804

Chlorine:
   #350(365)(14,690)/150,000 =   12,511   12,511

Diatomacious Earth:
   #60(365)(14,690)/75,000 =   4,289   4,289
                          Total   #67,634   #69,989

Sensitivity to interest rate:
PW at 20%   #59,925   #64,538
PW at 15% (from above)   67,634   69,989
PW at 10%   78,986   78,552

As we can see from these figures, increasing the i* will not change the indicated decision, but a reduction to 10% could change it.
Sensitivity to rate of growth of demand:
At an increase of 450 gal/day per year   69,962   72,282
 "   "   "   " 350 gal/day per year   67,634   69,989
 "   "   "   " 250 gal/day per year   65,308   67,698

The result is the conclusion that variation in the rate of growth in demand does not tend to reverse the decision.

PEE Solutions Manual

CHAPTER 15

Use of the Mathematics of Probability in Economy Studies

15-1
Expected annual cost of insurance plus fire damage.

	Sprinkler	No Sprinkler
Annual Ins. Premium	$66,960	$163,680
Expected Avg. Loss recovered (1/2 of above)	-33,480	-81,840
Expected Damage Suffered 3($12,400,000)(Ins. rate)(0.5)	100,440	245,520
	$133,920	$327,360

EUAC with Sprinkler system
= $185,600(A/P,30%,20) +$8,500 +$133,920
= $185,600(0.30159) +$8,500 +$133,920 = $198,395

Management might consider the possibility that a destructive fire, regardless of the insurance coverage, might cause the company to fail due to loss of customers, time required to rebuild, etc. The extreme event might be so frightening that a sprinkler system might be installed to reduce the probability of a fire of destructive proportions inspite of the outcome of the above analysis.

15-2
Insurance costs and expected values of loss are:

Annual costs:	Site A	Site B
Insurance premium	$9,750	$13,500
Expected loss recovery	-4,875	-6,750
Expected loss	12,188	16,875
Total	$17,063	$23,625

Difference (B-A) = $6,562

The owner can afford to pay as much as $6,562/0.20 = $32,810 more for Site A than for Site B assuming that the full price paid for the land can be recovered at any time.

15-3

Design	First Cost	CR Cost	Risk Cost	Total Annual Cost
A	$1,000	$70,440	$200,000	$270,440
B	1,125	79,245	125,000	204,245
C	1,500	105,660	50,000	155,660
D	1,750	123,270	25,000	148,270
E	1,960	138,062	12,500	150,562
F	2,400	169,056	5,000	174,056

(A/P,7%,75) = 0.07044

Plan D, a spillway to handle 14,500 cfs, is the most economical.
The solution is not sensitive to i in the range 5 to 9%, nor is it sensitive to a change of ± $500,000 in damages. Reducing the risk for Design E from 0.005 to 0.003, however, shifts the economic choice to Design E.

A question occasionally arises in class as to whether this model takes care of the probability of having a flood stage sufficient to exceed the spillway capacity more than once a year. That possibility should be included in the statement of probabilities. In other words, the probability given is supposed to be for a year, and if there is a real probability of an excessive flow occurring more than once, the probability should reflect it. The probability for a single event can be given for a month, a quarter, 6-month period, or a year. If the damage is the same each time an overflow occurs, then the probabilities can be added for the different periods in the year in order to obtain a probability for the year.

15-4

	Damages F(x)	Probability of stated damages, P(x)	E(x)
For Cap. of 1900 cfs	$ 0	(1.00 − 0.02)	$ 0
	250,000	(0.02 − 0.01)	2,500
	300,000	(0.01 − 0.005)	1,500
	400,000	(0.005 − 0.000)	2,000
			$6,000
For Cap. of 2,300 cfs	$ 0	(1.00 − 0.005)	$ 0
	250,000	(0.005 − 0.002)	750
	300,000	(0.002 − 0.001)	300
	400,000	(0.001 − 0.000)	400
			$1,450

15-5
(a) ($50,000/10)(P/A,15%,10) = $5,000(5.019) = $25,095

(b) ($50,000/10 −$50,000/50)(5.019) = $20,076

15-6
(a) ($500,000/10)(P/A,15%,10) = $50,000(5.019) = $250,950

(b) ($500,000/10 −$500,000/50)(5.019) = $200,760

If one believes that expected value is the correct basis for analysis and decision making, then the results would be treated the same. However, an average individual or small business concern probably would view the order of magnitude difference in bad outcomes with greater trepidation.

15-7
Since two different amounts of damage may occur if the spillway capacity is exceeded, different costs of risk must be computed for each design. For example, the probability of a flow return between 8,500 and 10,000 cfs for Design A is (0.08−0.05) = 0.03; that for a flow greater than 10,000 cfs is 0.05. Thus the risk cost for Design A is $1,750,000(0.03) +$3,750,000(0.05) = $52,500 +$187,500 = $240,000

Design	CR Cost	Risk Cost ($000 omitted)	Total
A	$70,440	$1,750(0.08−0.05) +$3,750(0.05)	$310,440
B	79,245	1,750(0.05−0.03) +3,750(0.03)	226,745
C	105,660	1,750(0.02−0.01) +3,750(0.01)	160,660
D	123,270	1,750(0.01−0.0075) +3,750(0.0075)	155,770
E	138,062	1,750(0.005−0.002) +3,750(0.002)	150,812
F	169,056	1,750(0.002−0.001) +3,750(0.001)	174,556

15-8
CR and Risk Costs for each alternative are shown in the solution to Problem 15-3.

Design	CR Cost	Risk Cost	Comparison	CR	Risk	B/C
A	$70,440	$200,000				
B	79,245	125,000	B vs A	$ 8,805	$75,000	8.52
C	105,660	50,000	C vs B	26,415	75,000	2.84
D	123,270	25,000	D vs C	17,610	25,000	1.42
E	138,062	12,500	E vs D	14,792	12,500	0.85
F	169,056	5,000	F vs D	45,786	20,000	0.44

The most economical choice is Design D. Its B/C ratio over the base case (A) is $175,000/$52,830 = 3.31

PEE Solutions Manual   Chapter 15

15-9
Expected value of annual cost of damage to orange crop

= $750,000(0.75)(4/50) = $45,000

EUAC to prevent crop loss =

$150,000(A/P,25%,10) +$9,500
= $150,000(0.28007) +$9,500 = $51,511

The sensitivity of the solution to changes in the various factors can be tested by solving the same problem with several different values. An increase in the selling price of oranges will increase the expected value of the damage due to frosts. An increase in the minimum attractive rate of return will increase the equivalent uniform cost of providing protection against frosts. An increase in the probability of "killing" frosts will increase the expected value of loss due to frosts.

The owner's decision might be influenced by financial ability to survive a killing frost (be able to meet mortgage payments, feed and house the family, etc.)

15-10
Expected annual cost = P(A/P,10%,50)
+$150,000(Probability of overtopping)

A. $162,500(0.10086) +$150,000(0.20) = $46,390
B. $21,181 +$15,000 = $36,181
C. $25,215 +$6,000 = $31,215
D. $31,267 +$3,000 = $34,267

15-11 (Data from Problem 15-10)

Design	CR Cost	Risk Cost	Comparison	CR	B	B/C
A	$16,390	$30,000				
B	21,181	15,000	B vs. A	$4,791	$15,000	3.13
C	25,215	6,000	C vs. B	4,034	9,000	2.23
D	31,267	3,000	D vs. C	6,052	3,000	0.50

Design C is most economical choice.

PEE Solutions Manual   Chapter 15

15-12
($000 omitted) in decision tree.

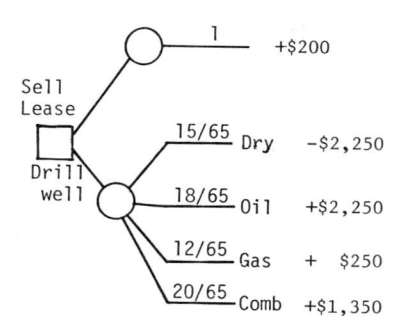

```
                        1
                       ○────── +$200
         Sell       ╱
         Lease    ╱
              ┌─┐    15/65
              │ │────────── Dry    -$2,250
              └─┘╲
              Drill╲  18/65
              well  ○────────── Oil    +$2,250
                   ╱╲ 12/65
                  ╱  ────────── Gas    + $250
                 ╱    20/65
                       ────────── Comb  +$1,350
```

Selling yields a certain $200,000
The expected net present value of drilling is +$565,385

15-13

NPW Payoff ($000 omitted)

	Dry S_1	Oil S_2	Gas S_3	Comb. S_4	minimum
Drill A_1	-2,250	2,250	250	1,350	-2,250
Sell A_2	200	200	200	200	200

The maximum security level strategy is to sell the lease for
$200,000. If independent oil companies such as that described in Problem
15-12 consistently applied this decision rule, they would seldom do any
drilling. Depending on the company's financial circumstances, it is
likely to view the problem from an entirely different perspective when a
site has a "track record" than when it does not. Previous successful
drillings in comparable circumstances will affect the decision-makers
intuition about the site.

15-14

$EPW(A_1) = \$560{,}000(0.6) + \$112{,}000(0.4) = \underline{\$380{,}800}$

$EPW(A_2) = \$1{,}040{,}000(0.6) - \$320{,}000(0.4) = \underline{\$496{,}000}$

$EPW(A_3) = [\$880{,}000(0.9) + \$80{,}000(0.1)](0.6)$

$\qquad -\$80{,}000(0.4) = \underline{\$448{,}000}$

15-15

(a) Minimums are: A_1, -36; A_2, -32; A_3, -33

The choice should be A_2 with the maximum of the minimums, $\$32{,}000$ annual cost.

(b) $EAC(A_1) = -\$26(0.3) - \$30(0.2) - \$24(0.4) - \$36(0.1) = \underline{-\$27.0}$

$EAC(A_2) = -\$17(0.3) - \$32(0.2) - \$30(0.4) - \$28(0.1) = \underline{-\$26.3}$

$EAC(A_3) = -\$33(0.3) - \$20(0.2) - \$22(0.4) - \$31(0.1) = \underline{-\$25.8}$

A_3 is preferred because it has the lowest expected annual cost, $\$25{,}800$.

15-16

Present worth of costs ($000 omitted):

$\$4{,}825 + \$475[(P/F, 20\%, 5) + (P/F, 20\%, 10)]$

$+ \$300(P/A, 20\%, 15) - \$3{,}000(P/F, 20\%, 15)$

$= \$4{,}825 + \$475(0.4019 + 0.1615) + \$300(4.675)$

$-\$3{,}000(0.0649) = \underline{\$6{,}300}$

Present worth of receipts:

$\$43(365)(150) = \$2{,}354{,}250/\text{yr.}$ at 100% occupancy

$PW(\$000 \text{ omitted}) = \$2{,}354(4.675) = \$11{,}006$

$EPW = \$11{,}006[0.8(0.3) + 0.6(0.6) + 0.35(0.1)]$

$\qquad - \$6{,}300 = \underline{\$689}$

The company must always consider alternative uses for its funds. The present worth with only 35% occupancy is $-\$2{,}448{,}000$ over the 15 years. It may wish to compare this figure with the low-end estimates of an alternative investment opportunity.

PEE Solutions Manual

CHAPTER 16

Aspects of Economy Studies For Government Activities

General Notes:
　　The analysis of project proposals presented by different government agencies has become a subject of careful Congressional study and debate. The Joint Committee on Economy in Government has held numerous hearings and investigations of practices, often focusing on the effects of the use of different interest rates by different agencies on the allocation of public funds. The desirability of using "opportunity cost" as the basis for setting interest rates to be used in project justification has been recognized and so reported to Congress by the Committee. Some executive directives have been issued calling for the use of interest rates of up to 10%, and considerable pressure has been put on all agencies to do a better job of applying sound economic principles to project proposal formulation and justification.
　　This chapter is very important for students whether they expect to work for government agencies or to be "just taxpayers." As interested citizens, their knowledge of proper analysis methods will enable them to influence government actions, and to demand better use of their tax dollars.
　　Good examples from local, state, and national project proposals can be found in the newspapers at almost any time. Students can be assigned special projects to look into the methods and interest rates used in justifying these projects. They can attend public hearings and ask intelligent questions, and then prepare reports to present to the class. The reality of the principles presented in this chapter should stimulate student interest and help prepare them for an active role in government decision processes.

16-1
　　(a) Overall B/C = $99,104/$46,853 = <u>2.12</u>
　　　　Incremental B/C ratios for:
　　(b) Flood Control　　　　= $10,975/$8,110 = <u>1.35</u>
　　(c) Irrigation　　　　　　= $54,875/$9,434 = <u>5.82</u>
　　(d) Power　　　　　　　　= $12,622/$9,992 = <u>1.26</u>
　　(e) Water　　　　　　　　= $16,462/$1,929 = <u>8.53</u>

16-2
　　B/C for single purpose projects:

　　　　(a) Flood Control:　　$10,975/$20,000 = <u>0.55</u>
　　　　(b) Irrigation:　　　　$54,875/$28,800 = <u>1.91</u>
　　　　(c) Water:　　　　　　$16,462/$15,600 = <u>1.06</u>

PEE Solutions Manual   Chapter 16

16-3
The "remaining justifiable expenditure" is the smaller of either benefits or alternative costs less the separable costs. The allocation of nonseparable costs is based on the percentage that the remaining justifiable expenditure for the individual purpose is of the total remaining justifiable expenditure. Therefore, it is not necessarily true that the purpose having the larger incremental B/C will <u>always</u> be allocated a larger percentage of the nonseparable costs. Under usual situations, however, the expectation is that a larger portion of the nonseparable costs will be allocated to the purpose with the larger incremental B/C simply because, the larger the benefits, the larger the alternative cost is likely to be.

The rational foundation for this method stems from the fact that most multipurpose projects grow out of an original single purpose project proposal. Commonly it is found that, by adding some additional investment to the original project, additional purposes can be served. A certain capacity dam might well serve for flood control purposes, but to add power generation and irrigation, additional expenditures must be made for increased height and strength, more land to be inundated, flumes and channels, etc. These additional expenditures cannot be entirely separated to the various purposes. The proposed method, in effect, says, "separate everything you can and allocate the remaining on the basis of the separate benefits to be enjoyed by the individual purposes." Basically this is the concept of basing charges on ability to pay.

16-4   (See table of Problem 16-5)
The capitalized costs and capitalized benefits based on 7% are shown in that table. Using values from that table, the B/C ratios are:

Total	;	$64,425/$44,280 = 1.45
Flood Control	;	$ 7,135/$ 8,114 = 0.88
Irrigation	;	$28,800/$ 8,812 = 3.27
Power	;	$ 8,205/$ 8,883 = 0.92
Water	;	$10,702/$ 2,014 = 5.31

The B/C ratios for Flood Control and Power are less than 1.0, thus these purposes will not pay for themselves.

16-5 (P/A,7%,100) = 14.269 ; ($000 omitted)

Item*	Flood Control	Irrigation	Power	Water	Fisheries	Navigation	Total
1							44,280
a							(39,500)
b							(4,780)
2	7,135	35,672	8,205	10,702	1,284	1,427	64,425
3	20,000	28,800	12,622	15,600	-	-	-
4	7,135	28,800	8,205	10,702	1,284	1,427	57,553
5	8,114	8,812	8,883	2,014	-	-	27,823
a	(8,000)	(6,800)	(6,700)	(1,600)	-	-	(23,100)
b	(114)	(2,012)	(2,183)	(414)	-	-	(4,723)
6	[-979]	19,988	[-678]	8,688	1,284	1,427	31,387
7	-	63.7%	-	27.7%	4.1%	4.5%	100%
8	-	10,483	-	4,559	674	741	16,457
a	-	(10,430)	-	(4,576)	(656)	(738)	(16,400)
b	-	(36)	-	(16)	(2)	(3)	(57)
9	8,114	19,295	8,883	6,573	674	741	44,280
a	(8,000)	(17,230)	(6,700)	(6,176)	(656)	(738)	(39,500)
b	(114)	(2,048)	(2,183)	(430)	(2)	(3)	(4,780)
10	8	141	153	29	2	2	335

*Items are same as in Table 16-1

Operation, maintenance, and replacement costs, as well as benefits, change substantially with a change in interest rate. Those purposes that have low annual costs but high construction costs have a greater reduction in their B/C ratio than those with low construction costs. In this problem, an increase in interest rate caused a decrease in all B/C ratios with two going below 1.0, Flood Control and Power. Consequently, these purposes are suspect and may be candidates to be eliminated.

The actual B/C ratios are given in Problem 16-4. Note that, in the allocation of joint costs, no allocation is made to those purposes which are unable to pay for themselves.

PEE Solutions Manual   Chapter 16

16-6
(a) Flood Control
If no nonseparable costs were allocated to flood control, the i at which B/C = 1 can be found by:

$500(P/A,i%,100) = $8,000 +$110(P/A,i%,100)/(P/A,4.5%,100)
$500(P/A,i%,100) = $8,000 +($110/21.950)(P/A,i%,100)
    (P/A,i%,100) = $8,000/$494.99 = 16.162
    From which i = **6.2%**

(b) With same assumption, let i = the break even interest rate for power:

$575(P/A,i%,100) = $6,700 +$3,292(P/A,i%,100)/(P/A,4.5%,100)
    (P/A,i%,100) = $6,700/$425.02 = 15.7639
    From which i = **6.36%**

16-7
(a) 4.5%, perpetual life:
B = $575/0.045 = $12,778; C = $1,600 +$329/(P/A,4.5%,100)(0.045)
  = $1,600 +$242 = $1,842
B/C = $12,778/$1,842 = **6.94**

(b) 4.5%, 50 year life:
B = $575(P/A,4.5%,50) = $11,363
C = $1,600 +$329(P/A,4.5%,50)/(P/A,4.5%,100)
  = $1,600 +$296 = $1,896
B/C = $11,363/$1,896 = **5.99**

(c) 4%, 100 year life:
B = $575(P/A,4%,100) = $14,090
C = $1,600 +$329(P/A,4%,100)/(P/A,4.5%,100) = $1,967
B/C = $14,090/$1,967 = **7.16**

(d) 5%, 100 year life:
B = $575(P/A,5%,100) = $11,413
C = $1,600 +$329(P/A,5%,100)/(P/A,4.5%,100) = $1,897
B/C = $11,413/$1,897 = **6.02**

16-8
All monetary figures in Table 16-1 are present worths. If the uniform annual worth figures had been used, all the monetary figures would have been multiplied by (A/P,4.5%,100) = 0.04556. No change would have occurred in the B/C ratios.

16-9
EUAC for government costs, Location X ($000 omitted)

Initial cost $6,776(.08)	=	$542.08
Resurface $160(11.3 mi)(A/F,8%,15) = $1,808(0.03683)	=	66.59
Maintenance $4(11.3 mi)	=	45.20
Total	=	$653.87

User costs, Location X

fuel [(0.15)($0.80) +(0.85)($0.216)](700)(11.3)(365)	=	$876,539
time 700(0.25)(11.3mi)(1/40mph)(60min/hr)($0.24)(365)	=	$259,844
Total		$1,136,383

EUAC for government costs, Location Y ($000 omitted)

Initial cost $4,640(0.08)	=	$371.20
Resurface $160(13.5 mi)(0.03683)	=	79.55
Maintenance $4(13.5 mi)	=	54.00
Total	=	$504.75

User costs, Location Y

fuel [(0.15)($0.80) +(0.85)($0.216)](700)(13.5)(365)	=	$1,047,192
time 700(0.25)(13.5 mi)(1/40mph)(60min/hr)($0.24)(365)	=	310,432
Total		$1,357,624

Incremental B/C ratio (X - Y):

$$B/C = \frac{\$1,357,624 - \$1,136,383}{\$653,870 - \$504,750} = \frac{\$221,241}{\$149,120} = \underline{1.48}$$

16-10
(Figures from solution to Problem 16-9. $000 omitted.)

EUAC's	Location X	Location Y
Government costs	$ 654	$ 505
User costs	1,136	1,358
	$1,790	$1,863

Location X is preferred by an EUAC of $73,000

16-11

EUACF(X) = -$6,776(i) -$1,808(A/F,i%,15) -$1,182

EUACF(Y) = -$4,640(i) -$2,160(A/F,i%,15) -$1,412

NAW(X-Y) = -$2,136(i) +$352(A/F,i%,15) +$230 = 0
from which i = <u>11.24%</u>. If i* is above 11.24%, Location Y is preferred.

PEE Solutions Manual   Chapter 16

16-12
   This is another possible problem for class discussion. A similar topic was used as an illustration in the introductory material regarding "benefits" near the beginning of Chapter 7.

16-13
   This is a possible class discussion problem to illustrate one type of conceptual difficulty that sometimes arises in the economic analysis of proposed public works. In the view of the authors, the $1.14 figure is the correct one to use rather than the $1.35 figure.
   If a public investment, such as a highway improvement, has the incidental effect of reducing the taxes paid by a portion of the public, such tax reduction should not be regarded as a project "benefit." Presumably, taxes are levied to meet the needs of governments, and the total amount of the taxes levied will be based on such needs as they appear to the people and their elected representatives. In the long run, a tax reduction in one place is likely to be made up by additional taxes collected in some other place.

16-14   ($000 omitted)
   Government costs (annualized):
   Investment: $2,520(A/P,7%,25) = $2,520(0.08581)   = $216.2
   Maintenance: 3.0
   Patrolling: -4.0
                                                           $215.2
   User impacts: Vehicle operating costs (increased)
     Trucks   17,000(0.20)($0.60)(0.10mi)(0.3)(365 days) = -$22.3
     Others   17,000(0.80)($0.15)(0.10mi)(0.3)(365 days) = - 22.3
                                                           -$44.6
   Time saved:
     Trucks & Comm: 17,000(0.40)($0.15)(0.30)(0.8)(365) = $ 89.4
     Passenger: 17,000(0.60)($0.05)(0.30)(0.8)(365) =   44.7
                                                           $134.1
   Accidents prevented:
     Fatal: (5/5)($150,000)(0.9)   = $135.0
     Non-fatal: (20/5)($30,000)(0.9)   = 108.0
     Property damage: (48/5)($2,000)(0.9)   =   17.3
                                                           $260.3

   B/C = (-$44.6 +$134.1 +$260.3)/$215.2 = **1.63**

16-15   (Cost calculations from solution to Problem 16-14)
   B/C = ($349.8 +$44.6 +$4.0)/($215.2 +$4.0 +$44.6)
       = $398.4/$263.8 = **1.51**
   The value of the ratio is reduced; however it remains greater than 1.

16-16
By eliminating passenger vehicle time savings:
B/C = ($349.8 -$44.7)/$215.2 = **1.42**
So long as accident reduction beneftis exceed $215.2 +$44.7 -$89.4 = $170.5(000) the project remains justified economically. Thus accident costs could be reduced ($260.3 -$170.5)/$260.3 = 34% and still justify the project. This problem can lead to much discussion of the "cost" of an accident and the value of life. Students should investigate what is done by the Department of Transportation (or whatever) in their state.

16-17
B/C = ($349.8 +$4.0 -$3.0)/$216.2 = **1.62**
There is no substantial impact on the B/C ratio in this case. It would be unusual for government annual disbursement figures to be large enough to change the B/C ratio substantially.

16-18   (P000 omitted)
1. Palmillas-Miquihana (3)   1,688 M/pesos per inhabitant served
2. Las Norias-Cruillas (1)   1,761   "   "   "   "
3. Mendez-Entronque (4)   1,763   "   "   "   "
4. El Capulin-Bustamante (2)   2,061   "   "   "   "

Because this is an _inverse_ cost effectiveness index, the lowest figure is the best and the priority list starts with Palmillas-Miquihana.

16-19
1.	Casas-Soto La Marina-La Pesca	(4)	2.11
2.	Llera-Gonzalez	(6)	2.02
3.	El Barretal-Padilla	(8)	2.01
4.	Mendez-Burgos	(10)	1.99
5.	El Barretal-Santa Engracia	(3)	1.82
6.	Altamira-Aldama	(2)	1.58
7.	Jimenez-Abasolo	(9)	1.50
8.	Ebano-Manuel	(7)	1.21
9.	Limon-Ocampo	(5)	1.11
10.	Hidalgo-La Mesa	(1)	0.90

16-20
Capitalized cost of Plan I (1/3 capacity now, 1/3 at 10 years, and 1/3 at 35 years):
$4,000,000 +$11,000/0.09 +[$4,000,000 +$11,000/0.09](P/F,9%,10)
   +[$4,000,000 +$11,000/0.09](P/F,9%,35)
= [$4,000,000 +$11,000/0.09][1 +0.4224 +0.0490] = $6,065,438

Plan II (1/2 capacity now, 1/2 at 20 years):
[$4,800,000 +$10,000/0.09][1 +(P/F,9%,20)]
= $4,911,111(1.1784) = $5,787,253

Plan III (2/3 capacity now, 1/3 at 35 years):
$5,700,000 +$9,500/0.09 +[$4,000,000 +$11,000/0.09](P/F,9%,35)
= $5,805,556 +$4,122,222(0.0490) = $6,007,544

Plan IV (full capacity now) = $6,800,000

Plan II is the most economical.

16-21
($000 omitted); (A/P,6%,50) = 0.06344
For LD compared to NFC:
   B = $600 -$225 = $375
   C = ($3,500 -$1,750)(0.06344) +$1,750(0.06) +$49.5 = $265.5
   B/C = **1.41**
For HD compared to LD:
   B = $225 -$20 -$15 = $190
   C = ($2,500 -$1,250)(0.06344) +$1,250(0.06) +$33.5 = $187.8
   B/C = **1.01**
The HD should be selected because the incremental B/C ratio for (HD-LD) is greater than 1. The overall B/C ratio for HD is:
   B = $600 -$20 -$15 = $565
   C = ($6,000 -$3,000)(0.06344) +$3,000(0.06) +$83 = 453.3
   B/C = **1.25**

16-22
($000 omitted); (A/P,10%,50) = 0.10086
For LD compared to NFC:
   B = $375; C = $1,750(0.10086) +$1,750(0.10) +$49.5 = $401
   B/C = **0.94**
For HD compared to NFC:
   B = $565; C = $3,000(0.10086) +$3,000(0.10) +$83 = $686
   B/C = **0.82**
Evaluated at an i* of 10%, neither project level is justified.

16-23 ($000 omitted); (A/P,6%,50) = 0.06344
For LD compared to NFC:
B = $375; C = $3,500(0.06344) +$49.5 = $271.5
B/C = <u>1.38</u>
For HD compared to LD:
B = $190; C = $2,500(0.06344) +$33.5 = $192.1
B/C = <u>0.99</u>
LD should be selected because the incremental B/C ratio for (HD-LD) is less than 1. However, the figures are so close that the final decision may be based on irreducibles. The overall B/C ratio for HD is:
B = $565; C = $6,000(0.06344) +$83 = $463.6 ; B/C = <u>1.22</u>

16-24 ($000 omitted); (A/P,9%,40) = 0.09296
174 ft. dam:
B = $120(19,450) = $2,334
C = $13,083(0.09296) +$280 = $1,496
B/C = <u>1.56</u>
194 ft. compared to 174 ft. dam:
B = $120(21,675 -19,450) = $267
C = ($14,460 -$13,083)(0.09296) +($320 -$280) = $168
B/C = <u>1.59</u>
211 ft. compared to 194 ft. dam:
B = $120(23,600 -21,675) = $231
C = ($16,450 -$14,460)(0.09296) +($390 -$320) = $255
B/C = <u>0.91</u>
The 194 ft. dam should be selected. Its overall B/C ratio is:
B/C = $2,601/$1,664 = <u>1.56</u>
The capacity cost of this power is $1,664,202/21,675 = $76.78 per kilowatt-year.

16-25
Attaching these additional costs to the project can only reduce the B/C ratios. If the 194 ft. choice from Problem 16-24 is not changed, then it alone needs to be reevaluated.
B = 0.9($267) = $240
C = $168(no change) ; B/C = <u>1.43</u>
The overall B/C ratio will be:
B = 0.9($2,601) = $2,341; C = $1,664 +$160(0.09296) +$6 = $1,685
B/C = <u>1.39</u> or $86.37 per kilowatt-year
Note that it is the electric power user who is expected to pay for these benefits rather than the beneficiaries themselves (e.g., through taxes or user charges). The class may wish to discuss the equity of such arrangements.

PEE Solutions Manual   Chapter 16

16-26 Daily water use is = 40,000(180) = 7.2 mgpd
Required plant capacity = 2(7.2) = 14.4 mgpd
Plant investment = 14.4($65,000) = $936,000
Annual bond redemption = $936,000/30 = $31,200/yr
First year interest = 0.07($936,000) =   65,520
Annual maintenance = 0.03($936,000) =   28,080
Annual chemicals = (320-70ppm)(7.2mgpd)($0.25)(365) =   164,250
Annual pumping = $3.60(7.2mgpd)(365) =   9,461
Annual labor =   36,000
      Total first year cost $334,511

EUAC @ 7% = $936,000(0.08059) +$237,791 = $313,223

Annual savings to citizens:
  soap = 40,000(38.5 -30.8)($0.90) = $277,200
  chemicals = $0.70(110)(320 -70 ppm) =   19,250
  heater replacement = $220(8,000)[(A/P,7%,8) -(A/P,7%,16)] =   108,434
      Total annual savings $404,884
B/C ratio = $405/$313 = **1.29**

The annual savings also exceed the first year cost of the system. If water rates are based on the first year cost, the necessary rate increase will be $334,511/7,200(365) = $0.1273 per 1,000 gal. If based on the EUAC, it will be $313,223/2,628,000 = $0.1192 per 1,000 gal.

A 7% interest rate was used both to value the investment cost and the user benefit. A discussion of both of these applications of interest charges would be in order. The two applications need not use the same value of i nor should they necessarily be based on the bond interest rate.

16-27 ($000 omitted)
     (A/P,8%,50) = 0.08174; (A/F,8%,50) = 0.00174
  A compared to NFC
     B = ($3,080 -$1,980) +$80 = $1,180
     C = $20,000(0.08174) -$1,000(0.00174) +$50 = $1,683
         B/C = **0.70**
  B compared to NFC
     B = ($3,080 -$700) +$120 = $2,500
     C = $23,000(0.08174) -$2,500(0.00174) +$70 = $1,946
         B/C = **1.28**
  C compared to B
     B = ($700 -$300) +$120 = $520
     C = $7,000(0.08174) -$1,000(0.00174) +$20 = $590
         B/C = **0.88**

Dam B should be chosen. It has an overall B/C ratio of 1.28 as shown because Dam A was rejected resulting in a comparison of Dam B with NFC.

PEE Solutions Manual

CHAPTER 17

Aspects of Economy Studies For Regulated Businesses

General Notes

Regulation has become so all pervasive that no business, regardless how small, can escape requirements to comply with some kind of government regulation often related to safety, health, or welfare. Many regulations can be met by any one of several alternative methods. This gives the business firm the opportunity to analyze the life-cycle cost of each alternative method before making a decision as to the method to be adopted. There are many examples of such decision opportunities, in engineering journals, the Wall Street Journal, and other publications, that the instructor can use to increase class interest in the application of engineering economy to regulatory matters. One case study with several related questions is included in the problems for this chapter.

Engineering economy problems relating to regulated public utilities, whether investor or government owned, are rather special and require a thorough knowledge of the rules and policies specified by the various regulatory bodies. It is suggested that the instructor make the examples more appealing to students by relating them to the local situation. Does the local regulatory agency use the concept of minimization of revenue requirements? Is it liberal or very tight regarding items included in the rate base and the permitted fair rate of return? Does it insist on flow-through of investment tax credits and other benefits of liberalized depreciation? All of these points can be used to help explain the application of engineering economy principles to the special problems of regulated industries. Since all persons are either willing or unwilling users of public utility services, students usually become interested in how their utility companies make investment decisions and how those decisions affect the rates they have to pay for service. This provides the opportunity to assign groups of students to study rate cases and investment proposals for independent project reports.

Giving students information regarding annual capital investments by utility companies in your area, and data on income and operating expenditures, as well as cases and hearings before regulatory bodies, will stimulate interest. Since utilities are so capital intensive, there are many opportunities for application of the principles of engineering economy to such widely different matters as proposed capital investments in plant expansion and whether or not the stock of a local utility company is an attractive investment.

PEE Solutions Manual   Chapter 17

17-1 Plan I
For land: n = 40; s = 100%; P = $10,000
           t = [0.40/0.60](0.11 -0.09(0.4))[1 +(1 -1)(1-(A/G,11%,40)/40)]
             = 0.66667(0.11 -0.0360)(1) = 0.0493333
For building and conduit: n = 40; s = 0; P = $200,000
           t = (0.40/0.6)(0.074)[0 +(1 -0)(1 -(A/G,11%,40)/40)]
             = 0.6667(0.074)(0.78835) = 0.038892
For cable: n = 20; s = 20%; P = $40,000
           t = (0.6667)(0.074)[(0.20 +(1 -0.20)(1 -(A/G,11%,20)/20)]
             = (0.6667)(0.074)(0.74964) = 0.036984
First installation:
       CR: $10,000(0.11) = $ 1,100
              $200,000(A/P,11%,40) =   22,344
              $40,000(A/P,11%,20) +$8,000(0.11) =   5,903
       EUACF for income taxes:
              $10,000(0.0493333) =   493
              $200,000(0.038892) =   7,778
              $40,000(0.036984) =   1,479
       Ad valorem tax = $250,000(0.03) =   7,500
       O & M disbursements   30,000
              Annual revenue required = $75,787
Second installation: n = 20; s = 20%; P = $60,000
       CR: $48,000(A/P,11%,20) +$12,000(0.11) = $7,348
       EUACF for income taxes: $60,000(0.036984) =   2,219
       Ad valorem taxes $60,000(.03) =   1,800
       O & M disbursements =   8,000
              Annual revenue required = $19,367
Third installation: same as second, except that it occurs 6 years later.

    Total annual revenue required for first 20 years:
      = $75,787 +$19,367(P/A,11%,12)(P/F,11%,8)(A/P,11%,20)
       +$19,367(P/A,11%,6)(P/F,11%,14)(A/P,11%,20)
      = $77,787 +$6,852 +$2,387 = $87,026

    Plan II
t's for land, building and conduit, and cable are the same
       CR: $10,000(0.11) = $ 1,100
              $200,000(A/P,11%,40) =   22,344
              $80,000(A/P,11%,20) +$20,000(0.11) =   12,246
       UACF for income taxes
              $10,000(0.0493333) =   493
              $200,000(0.038892) =   7,778
              $100,000(0.036984) =   3,698
       Ad valorem taxes = $310,000(0.03) =   9,300
       O & M disbursements =   38,000

              Total annual revenue requirement = $94,959

17-2 (a) Revenue required if leased: i = 12% (Since payments are monthly, continuous compounding is used for effective 12%)

```
Year
0-1      $4,800 x 0.9454 = $ 4,538
1-2       4,800 x 0.8441 =   4,052
2-3       3,360 x 0.7537 =   2,532
3-4       3,360 x 0.6729 =   2,261
                            $13,383
```

$13,383(A/P,12%,4) = **$4,406**

(b) Revenue required if purchased:
t = (0.40/0.60)(0.12 -0.10(0.50))[0.10 +0.90(1 -(A/G,12%,4)/4)]
  = 0.032440

Total revenue required annually = $18,000(A/P,12%,4) +$2,000(0.12)
  +$20,000(0.032440) = **$6,815**

(c) Revenue required if leased and in rate base:
t = (0.40/0.60)(0.12 -0.10(1.00))[0.10 +(0.90)(1 -(A/G,12%,4)/4)]
  = 0.00926

Total revenue required annually = $18,000(A/P,12%,4) +$2,000(0.12)
  +$20,000(0.00926) = **$6,351**

Comments: In the first case, the consumer receives the service without paying the company anything other than the bare cost of the service and without "paying" any income tax. If the truck is included in the rate base, the customer pays 12% on the beginning of year value, based on straight-line depreciation, and the income tax incurred by the company. With 100% debt financing, (c), the income tax is quite small, however it increases if the truck is only 50% financed by debt.
   It turns out that, under leasing without having the truck in the rate base, the consumer is paying less than straight-line depreciation ($18,000/4 = $4,500.) This means that the leasing company may have a "gimmick". Its price for the truck may be less than that for the utility company and it may expect to sell the used truck for much more than the $2,000 estimated by the utility company.
   Leasing without the item being in the rate base seems very unfair to a utility company because it is not compensated in any way for taking the risk involved in the acquisition of the item under a fixed lease-price contract.

PEE Solutions Manual   Chapter 17

17-3 (a) Imbedded cost of capital as of June 30, 1990
              20% x 0.065   = 0.0130
              20% x 0.090   = 0.0180
              30% x 0.135   = 0.0405
              30% x 0.105   = 0.0315
              Cost of capital = 0.1030 = **10.3%**

         (b) Imbedded cost of capital after July 1, 1990
              18.18% x 0.09   = 0.01636
              27.27% x 0.135   = 0.03682
              27.27% x 0.105   = 0.02864
              27.27% x 0.098   = 0.02673
              Cost of capital = 0.10855 = **10.855%**

17-4 (a) Cost of debt   0.10855 x 0.55 = 0.05970
              Cost of equity   0.12500 x 0.45 = 0.05625
              Fair rate of return =   0.11595

         (b) Current cost of new capital 0.098 x 0.55 = 0.05390
              Cost of equity   0.125 x 0.45 = 0.05625
              Fair rate of return =   0.11015

         (c) If the fair rate is set on current cost of capital, the company will need to reduce its service charges, but it will still have to pay the high interest rates on its imbedded debt (10.5% and 13.5%.) The market value of its stocks will probably decline. In times of inflation and rising interest rates, it helps utility companies if the fair rate of return is set on current cost of capital, but in periods of declining rates, the fair rate of return will be reduced, as seen in this problem.

PEE Solutions Manual   Chapter 17

17-5 (a) Change debt/equity ratio to 55/45. This only changes the income tax: (refer to the solution to Problem 17-1)
Land: t = (0.40/0.60)(0.11 -0.9(0.55))(1)
          = (0.6667)(0.06050)(1) = 0.040333
Building and conduit:
          t = (0.6667)(0.06050)(0.78835) = 0.031798
Cable:
          t = (0.6667)(0.06050)(0.74964) = 0.030237
    CR: $10,000(0.11) = $ 1,100
          $200,000(A/P,11%,40) =   22,344
          $80,000(A/P,11%,20) +$20,000(0.11) =   12,246
    UACF for income taxes:
          $10,000(0.040333)   403
          $200,000(0.031798)   6,360
          $100,000(0.030237)   3,024
    Ad valorem taxes   9,300
    O & M disbursements   38,000
          Total annual revenue requirement = $92,777

(b) Again pick up values from solution to Problem 17-1:
Land: t = (0.40/0.60)[0.10 -0.08(0.40)](1)
          = (0.6667)(0.06800)(1) = 0.045336
Building and conduit:
          t = (0.6667)(0.06800)[0 +(1 -0)(1 -(A/G,10%,40)/40)]
          = (0.6667)(0.06800)(0.77260) = 0.035438
Cable:
          t = (0.6667)(0.06800)[0.20+(1-0.20)(1-(A/G,10%,20)/20)]
          = (0.6667)(0.06800)(0.739680) = 0.033534
    CR: $10,000(0.10) = $ 1,100
          $200,000(A/P,10%,40) =   20,452
          $80,000(A/P,10%,20) +$20,000(0.10) =   11,397
    UACF for income taxes
          $10,000(0.045336) =   453
          $200,000(0.035438) =   7,088
          $100,000(0.033534) =   3,353
    Ad valorem taxes   9,300
    O & M disbursements   38,000
          Total annual revenue requirement = $91,143

(c) Increasing the debt-equity ratio with the same cost of debt and the same fair rate of return reduces the revenue required because the company pays out more in interest (a tax deductible expense) and has less taxable income on which to pay income tax. Decreasing the cost of debt and the fair rate of return means that the company must lower its service rates but also that it pays less interest. The effect is to reduce the annual revenue requirement.

PEE Solutions Manual   Chapter 17

17-6 n = 20; s = 0.613; P = $160,000; Only t for second
installation needs to be changed:
t = (0.40/0.60)(0.12-0.10(50))[0.613-(1-0.613)(1-(A/G,12%,20)/20)]
  = (0.6667)(0.07)[1-(1-0.613)(6.020/20)]
  = (0.6667)(0.07)(0.88351) = 0.04123

Plan B: First installation:
  CR: $112,000(A/P,12%,20) +$28,000(0.12) = $18,354
  O & M disbursements =   9,000
  Ad valorem taxes = (0.02)($140,000) =   2,800
  EUACF for income taxes = 0.0353($140,000) =   4,942
  Annual revenue requirement = $35,096
Second installation:
  CR: $128,000(A/P,12%,20) +$32,000(0.12) = $20,976
  O & M disbursements =   8,000
  Ad valorem taxes = (0.02)($160,000) =   3,200
  UACF for income taxes = (0.04123)($160,000) =   6,597
  Annual revenue requirement = $38,773
Total annual revenue requirement for Plan B
  = $35,096 +$38,773(P/A,12%,12)(P/F,12%,8)(A/P,12%,20)
  = $35,096 +$12,987 = **$48,083**

This method of dealing with the cotermination problem divides the second installation into two parts. The residual (or salvage) value is computed as the present worth at time 20 of the unrecovered investment. The use of this method slightly increases the revenue requirement for twenty years.

17-7
Compute a new t for the second installation with s = 0.52.
  t = (0.6667)(0.07)[(0.52 +(1 -0.52)(1 -(A/G,12%,20)/20)]
    = (0.6667)(0.07)(0.52 +0.33552) = 0.03993
  UACF for income taxes now is = $160,000(0.03993) = $6,388 compared with $5,648 in Example 17-3. Therefore the new annual revenue requirement for the second installation is: (use the necessary values from Example 17-3)
  $37,824 -$5,648 +$6,388 = $38,564

  Total annual revenue requirement
    = $35,098 +$38,564(P/A,12%,12)(P/F,12%,8)(A/P,12%,20)
    = $35,098 +$12,917 = **$48,815**
Again, the higher salvage value results in a slightly higher income tax rate and thus increases the total revenue requirement.

PEE Solutions Manual   Chapter 17

Comments on Problems 17-8 through 17-15
This case study did not specify what depreciation methods should be used now, nor the i*. Since the case is rather complex it is suggested that straight line depreciation be used, with a 20-year life for the building and chemical treatment plant and 10-year life for the plant equipment. Tax rates in effect at the time of working the problem should be used. The 1989 corporate rate of 34%, applicable between $75,000 and $100,000 taxable income, is used in these solutions.

Some preliminary calculations will be helpful regarding a retrofitted plant:

Depreciation, buildings = $250,000/30	=	$ 8,333
Depreciation, equipment = $125,000/12	=	10,417
President's salary	=	40,000
Unskilled labor = 10 x $10 x 2,000 hrs./yr.	=	160,000
Skilled labor = 4 x $15 x 2,000	=	120,000
Property taxes = $300,000(0.015)	=	4,500
Total		$343,250

Plant operating expenses, including materials, supplies, energy, maintenance, etc., = $600,000 -$343,250 -$90,000 = $167,750. These expenses are expected to remain the same except for the additional $5,000 a year for the chemical treatment plant operation.

Current value of existing plant and net realizable cash flow if sold now:

	Market Value	Book Value	Gain(Loss)
Land	$100,000	$ 50,000	$50,000
Building	50,000	125,000	(75,000)
Equipment	15,000	15,000	0

Since a corporation can only use long term losses to balance long term or short term capital gains, the net loss of $25,000 will not produce any tax advantage because there will be no long term gains until after the carry-forward time limit has expired.

Capital required for rehabilitation of existing plant:
    Building rehab. $190,000
    Chem treatment plant   50,000
    New equipment   140,000
    Total $380,000

Sources of capital:
    Disposal of old equipment $ 15,000
    Equipment fund   125,000
    Loan from bank   200,000
    Total $340,000

This is $40,000 short of the required investment. The owners need to know if the difference can be taken from working capital. Or can the loan be increased? Will each of the 5 owners invest an additional $8,000 in the company?

PEE Solutions Manual   Chapter 17

Depreciation charges for rehabilitated plant:
    Building: ($125,000 +$190,000)/20   = $15,750
    Equipment: $125,000/10   =   12,500
    Chemical treat. plant: $50,000/20   =   2,500
            Total $30,750

Terminal salvage value of plant and equipment 20 year hence:
    Land: $140,000 ($90,000 gain)
    Equipment: 15,000
            Total $155,000

Cash flow before debt service and income taxes:
    Income $600,000
    Less: Salaries and wages $320,000
          Property taxes   4,500
          Annual O & M   167,750
          Chem plant operation   5,000   497,250
          Net cash flow $102,750

Value of existing plant plus additional investments:
    Existing plant $165,000
    New investments   365,000

    Tax rates by brackets: (Since we are analyzing the whole company, we
must consider the different tax rates of the several brackets)
            State: Federal: Applicable for bracket
      First $50,000 @5%   First $50,000 @ 15%   19.3%
      Over $50,000 @8%   next $25,000 @ 25%   31.0%
                           next $25,000 @ 34%   39.3%
                           Over $100,000 @ 39%   43.9%

PEE Solutions Manual

Chapter 17

17-8

year	Cash flow before income taxes	Write-off of initial outlay for Tax purposes	Payment on debt	Payment of interest	Influence on taxable income	Influence of income taxes on cash flow	Cash flow after income taxes
0	($ 530,000)						
0	$ 200,000						($ 330,000)
1	$ 102,750	($ 30,750)	($ 40,000)	($ 24,000)	$ 48,000	($ 9,264)	$ 29,486
2	$ 102,750	($ 30,750)	($ 40,000)	($ 19,200)	$ 52,800	($ 10,518)	$ 33,032
3	$ 102,750	($ 30,750)	($ 40,000)	($ 14,400)	$ 57,600	($ 12,006)	$ 36,344
4	$ 102,750	($ 30,750)	($ 40,000)	($ 9,600)	$ 62,400	($ 13,494)	$ 39,656
5	$ 102,750	($ 30,750)	($ 40,000)	($ 4,800)	$ 67,200	($ 14,982)	$ 42,968
6	$ 102,750	($ 30,750)			$ 72,000	($ 16,470)	$ 86,280
7	$ 102,750	($ 30,750)			$ 72,000	($ 16,470)	$ 86,280
8	$ 102,750	($ 30,750)			$ 72,000	($ 16,470)	$ 86,280
9	$ 102,750	($ 30,750)			$ 72,000	($ 16,470)	$ 86,280
10	$ 102,750	($ 30,750)			$ 72,000	($ 16,470)	$ 86,280
11	$ 102,750	($ 30,750)			$ 72,000	($ 16,470)	$ 86,280
12	$ 102,750	($ 30,750)			$ 72,000	($ 16,470)	$ 86,280
13	$ 102,750	($ 30,750)			$ 72,000	($ 16,470)	$ 86,280
14	$ 102,750	($ 30,750)			$ 72,000	($ 16,470)	$ 86,280
15	$ 102,750	($ 30,750)			$ 72,000	($ 16,470)	$ 86,280
16	$ 102,750	($ 30,750)			$ 72,000	($ 16,470)	$ 86,280
17	$ 102,750	($ 30,750)			$ 72,000	($ 16,470)	$ 86,280
18	$ 102,750	($ 30,750)			$ 72,000	($ 16,470)	$ 86,280
19	$ 102,750	($ 30,750)			$ 72,000	($ 16,470)	$ 86,280
20	$ 102,750	($ 30,750)			$ 162,000	($ 54,443)	$ 138,307
20	$ 90,000					Total =	1197713
						NPV @16.7%	$ 982
						IRR equity=	16.7%

PEE Solutions Manual   Chapter 17

17-9
See Cash flow table in Problem 17-8. The equity in the existing plant, based on its resale value at this time plus the new investment less the $200,000 loan, gives a $330,000 equity in the renovated plant.
Find i, so that NPW = 0 = -$330,000 +$29,486(P/F,i%,1)
+$33,032(P/F,i%,2) +$36,344(P/F,i%,3) +$39,656(P/F,i%,4)
+$43,968(P/F,i%,5) +$86,280(P/A,i%,14)(P/F,i%,5) +$138,307(P/F,i%,20)
    from which i = **16.7%**

17-10
Examine the cash flow table of Problem 17-8. The net positive cash flow every year indicates that it can meet debt repayment plus interest with a good margin of safety.

17-11
Cash made available by liquidating existing plant:
Land	$100,000	Loss on building
Building	50,000	more than balances
Equipment	15,000	capital gain on land
Total	$165,000	

17-12
Compliance with regulatory requirements will increase equivalent uniform annual cost by:
Investment   Operation   Interest on loan
$240,000(A/P,20%,20) +$5,000 +[$24,000(P/A,20%,5)
    -$4,800(P/G,20%,5)](A/P,20%,20)
= $42,286 +$5,000 +$9,903 = **$64,189**

199

PEE Solutions Manual   Chapter 17

17-13
Preliminary calculations for Cash Flow Table for proposed new plant:

```
Tax rates:   State              Federal           Applicable for bracket
First $25,000 @3%     First $25,000    @15%              17.55%
Over $25,000 @5%      Next  $25,000    @15%              19.25%
                      Next  $25,000    @25%              28.75%
                      Next  $25,000    @34%              37.30%
                      Over $100,000    @39%              42.05%
```

Investment in plant:
```
Land                  $ 50,000                          S = $50,000
Building               250,000     n = 40 yrs.          S = 0
Chem. treat. plant      50,000     n = 20 yrs.          S = 0
Equipment              140,000     n = 10 yrs.          S = $15,000
     Total            $490,000
```

Depreciation:
```
Building = $250,000/40           =  $6,250
Chem. treat. plant = $50,000/20  =   2,500
Equipment = $125,000/10          =  12,500
     Total                         $21,250
```

Moving expenses deductible first year:
```
Skilled employees: 4 x $5,000                           $20,000
Moving files, tools, dies, etc.                          40,000
     Total deductible 1st. yr.                          $60,000
```

Cash flow before debt service and income taxes:
```
Income                                                 $600,000
Less:  Salaries and wages       $332,000
       Annual O & M              152,750
       Chem. plant operation       5,000                489,750
       Net cash flow (before adjustments)              $110,250
```

Adjustments:
```
Training expenses, first yr.    $13,333
Deductible moving, first yr.     60,000
     Net cash flow, year 1:                             $36,917
Property taxes after 5 yrs.       7,350
     Net cash flow, yrs. 6 to 40                       $102,900
```

Debt Service: Debt repayment = $50,000/yr. for 5 yrs.
Interest starts at $30,000 and decreases by $6,000 a year.

Cash flow table is on following page.

PEE Solutions Manual   Chapter 17

17-13 (cont)
Headings: (Same as Problem 17-8)

0	($ 490,000)						
0	$ 250,000						($ 240,000)
1	$ 36,917	($ 21,250)	($ 50,000)	($ 30,000)	($ 14,333)	$ 0	($ 43,083)
2	$ 110,250	($ 21,250)	($ 50,000)	($ 24,000)	$ 65,000	($ 9,392)	$ 26,858
3	$ 110,250	($ 21,250)	($ 50,000)	($ 18,000)	$ 71,000	($ 15,238)	$ 27,012
4	$ 110,250	($ 21,250)	($ 50,000)	($ 12,000)	$ 77,000	($ 17,134)	$ 31,116
5	$ 110,250	($ 21,250)	($ 50,000)	($ 6,000)	$ 83,000	($ 19,371)	$ 34,879
6	$ 102,900	($ 21,250)			$ 81,650	($ 18,868)	$ 84,032
7	$ 102,900	($ 21,250)			$ 81,650	($ 18,868)	$ 84,032
8	$ 102,900	($ 21,250)			$ 81,650	($ 18,868)	$ 84,032
9	$ 102,900	($ 21,250)			$ 81,650	($ 18,868)	$ 84,032
10	($ 22,100)	($ 21,250)			$ 81,650	($ 18,868)	($ 40,968)
11	$ 102,900	($ 21,250)			$ 81,650	($ 18,868)	$ 84,032
12	$ 102,900	($ 21,250)			$ 81,650	($ 18,868)	$ 84,032
13	$ 102,900	($ 21,250)			$ 81,650	($ 18,868)	$ 84,032
14	$ 102,900	($ 21,250)			$ 81,650	($ 18,868)	$ 84,032
15	$ 102,900	($ 21,250)			$ 81,650	($ 18,868)	$ 84,032
16	$ 102,900	($ 21,250)			$ 81,650	($ 18,868)	$ 84,032
17	$ 102,900	($ 21,250)			$ 81,650	($ 18,868)	$ 84,032
18	$ 102,900	($ 21,250)			$ 81,650	($ 18,868)	$ 84,032
19	$ 102,900	($ 21,250)			$ 81,650	($ 18,868)	$ 84,032
20	($ 22,100)	($ 21,250)			$ 81,650	($ 18,868)	($ 40,968)
21	$ 102,900	($ 21,250)			$ 81,650	($ 18,868)	$ 84,032
⋮	$ 102,900	($ 21,250)			$ 81,650	($ 18,868)	$ 84,032
37	$ 102,900	($ 21,250)			$ 81,650	($ 18,868)	$ 84,032
38	$ 102,900	($ 21,250)			$ 81,650	($ 18,868)	$ 84,032
39	$ 102,900	($ 21,250)			$ 81,650	($ 18,868)	$ 84,032
40	$ 102,900	($ 21,250)			$ 331,650	($ 103,868)	($ 968)
40	$ 315,000				NPV @	16.27%=	($ 102)

201

17-14
Total payroll = $332,000 per year; sales tax gains to state = $332,000(0.4)(0.05) = $6,640/year
Property taxes foregone for 5 years = $490,000(0.015) = $7,350 per year
Find i so that:
$6,640(P/A,i%,40) = $7,350(P/A,i%,5)
i is approximately 55%, an excellent investment for the state!

17-15
Students should use the same spreadsheet that was prepared for the solution to Problem 17-13. Have the computer calculate the NPW of the Cash Flow after Taxes column at 14%.

NPW @ 14% = $53,062 for 40 years
EUACF = $53,062(A/P,14%,40) = $7,468

That means that the project will earn $7,468 a year more than is necessary to earn 14% on the equity investment.

17-16
Applicable tax rate = 0.08 +(1 -0.08)(0.34) = 39.28%
Before-tax cash flows for bonds at 10% and 6%

Year	10%	6%	Difference	
0	+$500,000	+$500,000	0	
1-40	-$25,000	-$15,000	+$10,000	semi annually
40	-$500,000	-$500,000	0	

After-tax difference = $10,000(1 -0.3928) = $6,372 per half year
Subsidy = $6,372(P/A,6%,40) = $95,875

Since the company is a regulated public utility, the consumers will pay $12,744 less per year ($6,372 x 2) if the bonds are tax exempt at 6% than they would pay if the taxable bonds were issued at 10%. Stockholders will benefit to the extent that the fair rate of return is higher than the cost of capital included in the rate base.

PEE Solutions Manual   Chapter 17

17-17

Control Systems	A	B	C
Total annual disbursements	-$126,400	-$531,900	-$182,000
Depreciation	- 64,100	- 90,700	- 53,800
Total deductible expenses	-$190,500	-$622,600	-$235,800
Extra taxable income over A		-$432,100	-$ 45,300
Extra income tax over A		+$172,840	+$ 18,120
Cash flow after taxes	-$126,400	-$359,060	-$163,880
Capital recovery	^a -$130,572	^b -$135,170	^a -$109,591
EUACF	-$256,972	-$494,230	-$273,009

^a = P(A/P,8%,20) = 0.10185;   ^b = P(A/P,8%,10) = 0.14903

(b) Only change is in capital recovery: i = 14%
    EUACF(A) = -$319,964; EUACF(B) = -$536,944; EUACF(C) = -$326,340

(c) If all systems experienced the same variation in all costs (first cost as well as annual disbursements), there would be no change in the attractiveness of the three systems.

On the other hand, if the costs of B should be lower by 30% and the cost of A should increase by 30%, then there would be very little difference between the two systems as regards EUACF.

It is likely that the estimates of first cost will be more accurate than the estimates of operating expenditures. A sensitivity analysis could be performed to see how sensitive the results are to variations (pessimistic and optimistic) in the operating costs.

PEE Solutions Manual

APPENDIX A

Continuous Compounding of Interest and the Uniform-Flow Convention

General Notes

Most students are familiar with open end charge accounts, borrowing from banks, financing cars, and savings accounts, yet few have ever considered the true cost of borrowed money, or the true rate of return on their savings accounts. The personal approach of some problems will help get students interested in the concepts of continuous compounding and the uniform-flow concept.

Many business firms tend to disburse funds more or less uniformly throughout the year and to collect their revenues fairly uniformly as well. In that case, the uniform-flow convention may better represent the real world situation than the assumption of year-end payments and receipts. However, it is not appropriate to assume uniform-flow for all cash flows. (See Problem A-3.) In engineering economy studies, most of the data employed are estimates of future events. As such, they probably will be in error by a much larger degree than the difference between the year-end convention and the uniform-flow convention. One should guard against assuming greater confidence in the resulting analysis because the uniform-flow convention has been used.

A-1 (a) let $r = 0.0575$; $e^r - 1 = 1.059185 - 1$
$i_{eff} = 5.9185\%$

(b) $i = (1 + r/365)^{365} - 1 = 0.059180$
$i_{eff} = 5.9180\%$

The difference between continuous and daily compounding only occurs in the 6th decimal place when the interest is expressed as a decimal fraction.

A-2
(a) $i = (1 + 0.015)^{12} - 1 = \underline{19.562\%}$ This is the effective annual rate of 1.5%/month, compounded monthly.

(b) $i = e^r - 1 = e^{0.18} - 1 = \underline{19.722\%}$ Using the continuous compounding result, 19.722%, would be closer to the truth than using 18% APR.

PEE Solutions Manual   Appendix A

A-3 Annual receipts and disbursements:
            Receipts: 52($185) = $9,620
            Disbursements: -12($520) =   - 6,240
            Annual taxes: -   240
                   Net cash flow =   +$3,140

   (a) End-of-year convention:
       NPW = -$6,000 +$3,140(P/A,25%,6) = +$3,266.14

   (b) Uniform-flow convention:
       NPW = -$6,000 +$3,380(P/Ā,25%,6) -$240(P/A,25%,6)
           = -$6,000 +$3,380(3.307) -$240(2.951) = +$4,469.42

   (c) Percentage difference: ($4,469.42 -$3,266.14)/$3,266.14
       = 0.3683 or 36.83%. This is a very substantial difference,
       which is due primarily to the high interest rate required
       before income tax.

A-4   Difference in
                                                                Interest
   (a) $1,000(1.12) = $1,120

   (b) $1,000(1 +0.06)^2 = $1,123.60   3.60%

   (c) $1,000(1 +0.03)^4 = $1,125.50   1.91%

   (d) $1,000(1 +0.01)^{12} = $1,126.83   1.32%

   (e) $1,000(1 +0.12/52)^{52} = $1,127.34   0.51%

   (f) $1,000(1 +0.12/365)^{365} = $1,127.47   0.13%

   (g) $1,000(e^{0.12})$ = $1,127.50   0.03%

As the frequency of compounding increases, the rate of increase in
the effective annual interest decreases. Going from annual compounding
to monthly compounding increases the effective rate by 6.83%, but from
monthly to continuous compounding only increases the rate by 0.67%.

A-5
   (a) NPW = -$25,000 +$8,500(0.885 +0.689 +0.537 +0.418 +0.326)
           = -$25,000 +$8,500(2.855) = -$25,000 +$24,267 = -$732

   (b) NPW = -$25,000 +$8,500(0.8963) +0.7170 +0.5736 +0.4589 +0.3671)
           = -$25,000 +$8,500(3.0129) = -$25,000 +$25,510 = +$613

   (c) The lower value in (a) is due to discounting with a higher
interest rate. The difference is that in (a), the actual effective
interest rate is greater than 25%, whereas, Table D-30 is based on a
lower "force of interest" in order to make the interest factors yield an
effective interest rate of exactly 25%.

PEE Solutions Manual   Appendix A

A-6
(a) Find i, so that NPW = 0 = -$500(F/$\overline{P}$,i%,1) +$110(P/$\overline{A}$,i%,10)
    +$20(P/F,i%,10)
    = -$500(P/$\overline{F}$,i%,1)(1 +i) +$110(P/$\overline{A}$,i%,10) +$20(P/F,i%,10)
    at 15%: NPW = -$500(0.933)(1.15) +$110(5.386) +$20(0.2472)
                = +$60.93
    at 20%: NPW = -$500(0.914)(1.20) +$110(4.599) +$20(0.1615)
                = -$39.28
        i is approximately <u>18%</u>

(b) Find i, so that NPW = 0 = -$500(F/P,i%,1) +$110(P/A,i%,10)
    +$20(P/F,i%,10)
        i is approximately <u>14.3%</u>

A-7
(a) Uniform-flow convention:
    Proposal M: NPW = 0 = -$75,000 +$17,200(P/$\overline{A}$,i%,6)

                 (P/$\overline{A}$,i%,6) = 4.3605; i = <u>11.93%</u>

    Proposal N: NPW = 0 = -$75,000 +$11,700(P/$\overline{A}$,i%,10)
                            +$15,000(P/F,i%,10)
        By interpolation between 10% and 12%, i = <u>11.95%</u>

(b) End-of-year convention:
    Proposal M: NPW = 0 = -$75,000 +$17,200(P/A,i%,6)
                 (P/A,i%,6) = 4.3605;   i = <u>9.96%</u>

    Proposal N: NPW = 0 = -$75,000 +$11,700(P/A,i%,10)
                            +$15,000(P/F,i%,10);   i = <u>10.76%</u>

A-8
Proposal M: There will be no change in the prospective rate of
    return by either method.
Proposal N: NPW = 0 = -$75,000(P/$\overline{F}$,i%,1)(1 + i)
    +$11,700(P/$\overline{A}$,i%,10) +$15,000(P/F,i%,10)
    @ i = 10%, NPW = +$2,507.4
    @ i = 12%, NPW = -$4,548.9;   i = <u>10.71%</u>

No changes will be made in the rates of return by the end-of-year
method.
The fact that the machine involved in Proposal N will be built during
the entire year prior to operation reduces i from 11.95% to 10.71% using
the uniform-flow convention.

PEE Solutions Manual   Appendix A

A-9

(a) $i = (1 + 0.0525/365)^{365} - 1 = 0.053899$ or <u>5.3899%</u>

(b) $i = e^{0.0525} - 1 = 0.053903$ or <u>5.3903%</u>

(c) Deposits can be made or funds can be withdrawn at any time during the working day. Daily compounding eliminates arguments about interest due. Furthermore, continuous compounding is even less well understood by the general public than is periodic compounding.

A-10

End-of-year convention: ($000 omitted)
NPW = 0 = -$20,000(F/P,i%,1) -$10,000 +$2,700(P/A,i%,40)
          -$1,000(P/A,i%,10) +$100(P/G,i%,10)
@ i = 7%, NPW = +$343.6
@ i = 8%, NPW = -$3,515.9; i = <u>7.08%</u>

Uniform-flow convention: ($000 omitted)
No tables are given for converting $100,000 gradient, flowing uniformly throughout the year, to a PW. Thus, the $100,000 must be converted to an end-of-year gradient by $100,000(P/$\overline{F}$,i%,1)(1+i). Then the end-of-year factors can be used.

NPW = 0 = -$20,000(P/$\overline{F}$,i%,1)(F/P,i%,2) -$10,000(P/$\overline{F}$,i%,1)(F/P,i%,1)

          +$2,700(P/$\overline{A}$,i%,40) +$100(P/G,i%,10)(P/$\overline{F}$,i%,1)(F/P,i%,1)

          -$1,000(P/$\overline{A}$,i%,10)
@ i = 7%, PW = +$355.6
@ i = 8%, PW = -$3,656.4; from which i = <u>7.09%</u>

With very large amounts of money involved and high interest rates, the use of the uniform-flow convention can make a real difference. In this case, while the uniform-flow assumption is more realistic than the end-of-year assumption, it makes very little difference in the results obtained because of the long planning horizon.

PEE Solutions Manual   Appendix A

A-11
By end-of-year convention, revenue requirement
= $1,000,000/$5,000,000 = .20 or 20% of rate base

By uniform-flow convention, rate base becomes

= $2,500,000(P/$\overline{F}$,10%,1)(F/P,10%,2)

+$2,500,000(P/$\overline{F}$,10%,1)(F/P,10%,1)
= $2,500,000(0.9538)(1.21 +1.10) = $5,508,195

If revenue requirement is based on uniform-flow:
  with no change in rate base: Annual revenue requirement

= $1,000,000($\overline{A}$/F,10%,1) = $1,000,000(0.9531) = $953,100

with rate base computed by uniform-flow method: Annual

revenue requirement = (0.20)($5,508,195)($\overline{A}$/F,10%,1) = $1,049,974

Note that the ($\overline{A}$/F,10%,1) factor must be calculated from the formula on page 488 of the text.

PEE Solutions Manual

APPENDIX B

Cash Flow Series with Two or More Reversals of Sign

B-1

An analyst should first consider the purchase of the lease assuming that water flooding will not be undertaken. The cash flow series to be evaluated is:

Lease Purchase		
Yr.	Cash Flow	
0	-$180	
1	+ 120	
2	+ 90	
3	+ 60	
4	+ 30	

It is evident from inspection that the prospective rate of return is relatively high. Trial and error calculations indicate that the rate is approximately 31.5%. Thus, by normal standards, the investment should be made. (Unless i^* is greater than 31.5%.) Now the analyst can evaluate the attractiveness of water flooding.

Water Flooding	
Yr.	Cash Flow
0	-$1,810
1	+ 600
2	+ 500
3	+ 400
4	+ 300
5	+ 200
6	+ 100

Since water flooding is not required, that possible investment can be analyzed separately. Trial and error calculations of the cash flow resulting from flooding indicate that the investment will yield only a 6% rate of return. The investment in water flooding is not nearly as attractive as the original investment in the lease.

B-2

This problem changes the cash flow at time zero from 0 to -$180,000, and results in there now being three reversals of sign. It is still evident that most of the cash flow takes place in years 5 through 11. One approach would be to assume an auxiliary rate and apply it to all of the cash flows from year zero through year 4, and compute the prospective rate of the return on the investment portion of the -$1,810,000 cash flow at year 5.

If the auxiliary rate is assumed to be 0% (an extreme value), X will be computed to be +$120, and X-$1,810 = -$1,690. On that basis, the prospective rate of return is about 8.8%. If a high rate, say 15%, is chosen for the auxiliary rate, the new X will be +$98.6, and X-$1,810 = -$1,711.4. On that basis the prospective rate of return is about 8.3%.

The three reversals of sign could be interpreted as first an investing period, followed by a borrowing period, and finally an investing period. Since the major cash flows occur in the last period, years 5 through 11, two auxiliary rates could be used. The first could be the i^* on the original investment at time zero. If that is assumed to be 20%, the Value of X at year 5 based on a borrowing rate of 10%, would be +$118. Then the prospective rate of return on the X -$1,810 = -$1,692 would be about 7.5%.

PEE Solutions Manual   Appendix B

B-3 ($000 omitted)
Auxiliary rate = 0%:   X = $120 +$90 +$60 +$30 = $300
         -$1,810 + X = -$1,510
         NPW = 0 = -$1,510 +$600(P/A,i%,6) -$100(P/G,i%,6)
    @ i = 12%, NPW = +$64; @ i = 15%, NPW = -$34; i = <u>14.0%</u>

Auxiliary rate = 5%:   X = $120 -$30(A/G,5%,5)(F/A,5%,5) = $348
         -$1,810 + X = -$1,462
         NPW = 0 = -$1,462 +$600(P/A,i%,6) -$100(P/G,i%,6)
    @ i = 15%, NPW = +$14; @ i = 20%, NPW = -$124; i = <u>15.5%</u>

Auxiliary rate = 10%: X = $120 -$30(A/G,10%,5)(F/A,10%,5) = $401
         - $1,810 + X = -$1,409
         NPW = 0 = -$1,409 +$600(P/A,i%,6) -$100(P/G,i%,6)
    @ i = 15%, NPW = +$67; @ i = 20%, NPW = -$71; i = <u>17.4%</u>

B-4 ($000 omitted)
Auxiliary rate = 0%:   Let year 4 be new 0 date:
Yr.	Cash Flow
0	+$100
1 to 6	+200/yr.
7 to 16	+100/yr.
17	-3,000

         NPW = 0 = $100 +$100(P/A,i%,16) +$100(P/A,i%,6)
                   -$3,000(P/F,i%,17)
         @ i = 2.5%, NPW = -$15.3
         @ i = 3%,   NPW = +$82.8;   i = <u>2.6%</u>

Auxiliary rate = 10%:
Available positive cash flow for year 5 = -$700(F/P,10%,5)
     +$200(F/A,10%,5) = +$93.7
Yr.	Cash Flow
0	+$93.7
1 to 5	+200/yr.
6 to 15	+100/yr.
16	-3,000

         NPW = 0 = +$93.7 +$100(P/A,i%,15)
                   +$100(P/A,i%,5) -$3,000(P/F,i%,16)
         @ i = 4%, NPW = +$49.0
         @ i = 3.5%, NPW = -$33.2; i = <u>3.7%</u>

Auxiliary rate = 20%: Let year 7 be 0 date:
Yr.	Cash Flow
0	+$75
1 to 3	+200/yr.
4 to 13	+100/yr.
14	-3,000

Available positive cash flow for year 7 =
     -$700(F/P,20%,7) +$200(F/A,20%,7) = +$75
NPW = 0 = +$75 +$100(P/A,i%,13) +$100(P/A,i%,3)
          -$3,000(P/F,i%,14)
@ i = 6%, NPW = -$99.3
@ i = 7%, NPW = +$ 9.8;   i = <u>6.9%</u>

PEE Solutions Manual   Appendix B

B-5

Year	Net cash flow (and PW at 0%)	2.0%	6.0%	Present worth at 10.0%	15.0%	20.0%	30.0%	40.0%	50.0%
0	-180000	-180000	-180000	-180000	-180000	-180000	-180000	-180000	-180000
1	100000	98039	94340	90909	86957	83333	76923	71429	66667
2	100000	96117	89000	82645	75614	69444	59172	51020	44444
3	100000	94232	83962	75131	65752	57870	45517	36443	29630
4	100000	92385	79209	68301	57175	48225	35013	26031	19753
5	100000	90573	74726	62092	49718	40188	26933	18593	13169
6	-100000	-88797	-70496	-56447	-43233	-33490	-20718	-13281	-8779
7	-100000	-87056	-66506	-51316	-37594	-27908	-15937	-9486	-5853
8	-100000	-85349	-62741	-46651	-32690	-23257	-12259	-6776	-3902
9	-100000	-83676	-59190	-42410	-28426	-19381	-9430	-4840	-2601
10	-100000	-82035	-55839	-38554	-24718	-16151	-7254	-3457	-1734
11	0	0	0	0	0	0	0	0	0
12	0	0	0	0	0	0	0	0	0
13	0	0	0	0	0	0	0	0	0
14	0	0	0	0	0	0	0	0	0
15	0	0	0	0	0	0	0	0	0
16	0	0	0	0	0	0	0	0	0
17	0	0	0	0	0	0	0	0	0
18	0	0	0	0	0	0	0	0	0
19	0	0	0	0	0	0	0	0	0
20	200000	134594	62361	29729	12220	5217	1052	239	60
Sum PW	20000	-972	-11175	-6571	774	4092	-988	-14085	-29146
Rate of return		1.9%	1.9%	1.9%	29.0%	14.3%	14.3%	29.0%	29.0%	29.0%

PEE Solutions Manual   Appendix B

B-5 (cont)
Lotus 1-2-3 plot of NPW versus interest rate, Example B-3.

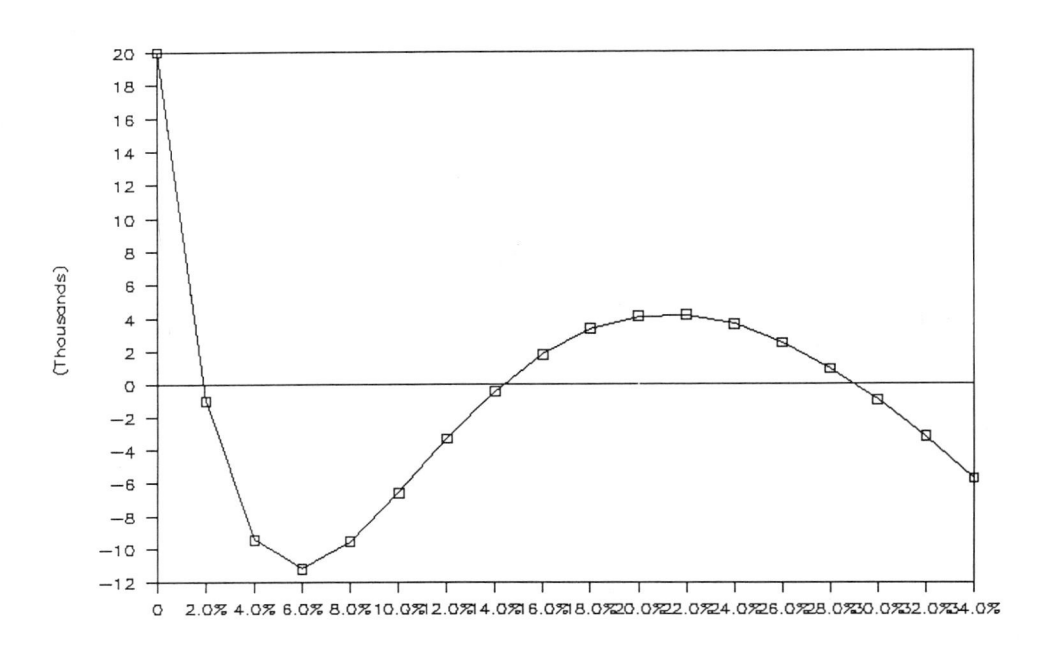

B-6
Plot the values of the NPW for different interest rates found by:
    NPW = -$16,000 +$50,000(P/A,i%,2) -$100,000(P/F,i%,3)
Values are:

i%	NPW	i%	NPW
0	-$16,000	200	+$2,518
10	- 4,355	250	+ 35
15	- 466	275	- 1,007
50	+ 9,926	300	- 1,936
100	+ 9,000		

Approximate solving values of i are 15.7% and 250.8%.

212

PEE Solutions Manual   Appendix B

B-7
Year 1 is the new zero date for all auxiliary rates.
Auxiliary rate (investing) = 0%; Cash available = $34,000
NPW = 0 = +$34,000 +$50,000(P/F,i%,1) -$100,000(P/F,i%,2)
         i (borrowing cost) = <u>13.06%</u>

Auxiliary rate = 10%; Cash available = $50,000
  -$16,000(F/P,10%,1) = $32,400
NPW = 0 = +$32,400 +$50,000(P/F,i%,1) -$100,000(P/F,i%,2)
         i = <u>14.72%</u>

Auxiliary rate = 15%; Cash available = $50,000
  -$16,000(F/P,15%,1) = $31,600
NPW = 0 = +$31,600 +$50,000(P/F,i%,1) -$100,000(P/F,i%,2)
         i = <u>15.58%</u>

B-8
(a)

i%	NPW
0	-$20,000
10	- 1,119.73
15	+ 2,700.42
20	+ 4,380.58
40	+ 938.56
45	- 995.51
50	- 3,059.01

Computer solution: Two solving rates: i = 11.46% and i = 41.75%

Solutions by interpolation between 10% and 15% and between 40% and 45% will be only slightly different.

(b) This is an investing problem, because the major flows occur immediately after the primary investment. The borrowing period occurs at the end of the series, so one must work backwards from the end to find out which positive cash flows are the returns on the initial investment.
   PW = -$60,000(P/F,10%,1) -$50,000(P/F,10%,2) -$40,000(P/F,10%,3)
      = -$125,920 at end of year 3
        $125,920 -$90,000 = $35,920 to be applied to year
           earlier receipt
        $50,000 -$35,920(P/F,10%,1) = $17,345
   NPW = 0 = -$70,000 +$60,000(P/F,i%,1) +$17,345(P/F,i%,2)
            i = <u>8.54%</u>

With this low rate of return, it is easy to see why the company resisted the government's pressure to withdraw the gas earlier. The company's i* for investments in proved fields was about 20% after income tax.

213

PEE Solutions Manual   Appendix B

B-9
(a) Taking the difference in the two cash flows, we have:
NPW = 0 = -$30,000 +$70,000(P/F,i%,20)
          -$400(P/A,i%,20)(P/F,i%,20)
   from which i = <u>3.9%</u>
   This indicates that the investment in the longer-lived structure is not very attractive.

(b) The NPW of this series of cash flows will become more negative as i is increased above 3.9% because of the large +$70,000 cash flow at year 20, the value of which diminishes rapidly as the interest rate is increased. The only other solving rate for this series is negative and therefore meaningless.

(c) The PW of the negative cash flows of $400 a year can be considered to be repayment of a sum borrowed at year 20. If a reasonable rate for borrowed money of 12% is used, the positive cash flow at year 20 resulting from the investment of $30,000 at year 0 is:
   $70,000 -$400(P/A,12%,20) = +$67,012
   NPW = 0 = -$30,000 +$67,012(P/F,i%,20); i = <u>4.1%</u>
   Clearly this is not an attractive investment.

Comments on Problems B-10 and B-11

In addition to being two good exercises in the use of spreadsheets and program graphics capabilities, these problems illustrate two important points.

For spreadsheet programs requiring a "seed" value of i to solve for rate of return, that solving rate may be <u>one</u> of the actual solving rates but there is no way to tell <u>which</u> one. When the value of i used to find the NPW was used as the seed, different solving values of i were obtained with these different seeds. Interestingly enough, all positive real roots were found over the spectrum of values of i used to find the NPW's. (Also see the spreadsheet for Problem B-5.) The spreadsheet will not tell you that this result is a distinct possibility. As the analyst, you must be well enough informed to know that more than one solving rate is possible and that neither/none may be meaningful.

For the data of Table B-1, the two solving roots are 26.7% and 37.1%. If the year 1 value is changed from +$120,000 to +$130,000, there is no real positive solving value of i. That is, the curve of NPW, which is concave upward throughout, never reaches 0. In Table B-2, the two solving roots are 2.8% and 26.2%. If the year 0 value is changed from -$700,000 to -$1,100,000, there is no real positive solving root. The curve of NPW, which is convex upward throughout, never reaches 0. It is hard to tell what "solution" (i.e., solving rate of return) any particular program will "find", if it "finds" any. Lotus 1-2-3 gives simply "ERR" in the cases just discussed without explaining what the "error" is.

These examples illustrate clearly the caveats expressed on pages 124-125 of the text concerning the use of spreadsheets.

PEE Solutions Manual   Appendix B

B-10 Sample spreadsheet for Table B-1.

Year	Net cash flow (and PW at 0%)	5.0%	10.0%	Present worth at 15.0%	20.0%	25.0%	30.0%	35.0%	40.0%
1	120	114	109	104	100	96	92	89	86
2	90	82	74	68	63	58	53	49	46
3	60	52	45	39	35	31	27	24	22
4	30	25	20	17	14	12	11	9	8
5	-1810	-1418	-1124	-900	-727	-593	-487	-404	-337
6	600	448	339	259	201	157	124	99	80
7	500	355	257	188	140	105	80	61	47
8	400	271	187	131	93	67	49	36	27
9	300	193	127	85	58	40	28	20	15
10	200	123	77	49	32	21	15	10	7
11	100	58	35	21	13	9	6	4	2
Sum PW	590	303	146	63	22	3	-3	-2	3
Rate of return		37.1%	37.1%	26.7%	26.7%	26.7%	26.7%	26.7%	37.1%	37.1%

PEE Solutions Manual   Appendix B

B-11

EXAMPLE B-2; TABLE B-2

Year	Net cash flow (and PW at 0%)	2.0%	3.0%	Present worth at 7.0%	10.0%	15.0%	20.0%	25.0%	30.0%
0	-700	-700	-700	-700	-700	-700	-700	-700	-700
1	200	196	194	187	182	174	167	160	154
2	200	192	189	175	165	151	139	128	118
3	200	188	183	163	150	132	116	102	91
4	200	185	178	153	137	114	96	82	70
5	200	181	173	143	124	99	80	66	54
6	200	178	167	133	113	86	67	52	41
7	200	174	163	125	103	75	56	42	32
8	200	171	158	116	93	65	47	34	25
9	200	167	153	109	85	57	39	27	19
10	200	164	149	102	77	49	32	21	15
11	100	80	72	48	35	21	13	9	6
12	100	79	70	44	32	19	11	7	4
13	100	77	68	41	29	16	9	5	3
14	100	76	66	39	26	14	8	4	3
15	100	74	64	36	24	12	6	4	2
16	100	73	62	34	22	11	5	3	2
17	100	71	61	32	20	9	5	2	1
18	100	70	59	30	18	8	4	2	1
19	100	69	57	28	16	7	3	1	1
20	100	67	55	26	15	6	3	1	1
21	-3000	-1979	-1613	-725	-405	-159	-65	-28	-12
Sum PW	-700	-146	28	337	360	268	141	25	-71
Rate of return		2.8%	2.8%	2.8%	2.8%	2.8%	ERR	26.2%	26.2%

PEE Solutions Manual

APPENDIX C

The Reinvestment Fallacy in Project Evaluation

General Notes:

Occasionally, the literature of capital investment analysis contains articles suggesting the use of methodologies that employ two or more interest rates in compound interest calculations carried out over the same period of years. The problems in this Appendix are intended to demonstrate two important points. First, improper inferences, and possibly errors in economic choice, can result any time two interest rates are employed over the same span of years. Second, the decision to invest in a project is separable and distinct from the decision concerning the disposition of recovered funds. As we have explained repeatedly, separable decisions should be made separately. The application of a methodology that is based on muddled thinking cannot be justified simply on the basis that it is easy to apply.

C-1

(a) ROR after income taxes:
NPW = 0 = -$75,700 +$20,000(P/A,i%,20) -$1,000(P/G,i%,20)
by interpolation, <u>i = 20.0%</u>

(b) Compound amount of the positive cash flow in 20 years at 3%:
[$20,000(P/A,3%,20) -$1,000(P/G,3%,20)](F/P,3%,20) = $308,395
At what interest rate must $75,700 now be compounded to amount to $308,395 in 20 years?
(F/P,i%,20) = $308,395/$75,700 = 4.074; <u>i = 7.3%</u>

(c) The i found in (a) is the interest rate at which the original investment will be returned to the investor, but in (b), the i of 7.3% is a combination of the 20.0% earned on the project and 3% earned elsewhere on the cash throw-off from the initial investment. Thus the earnings of one investment are being improperly influenced by a separable decision, that being the disposition of funds recovered from the project.

PEE Solutions Manual   Appendix C

C-2

(a) ROR after income taxes:
NPW = 0 = -$170,750 +$20,000(P/A,i%,20) -$1,000(P/G,i%,20)
i = 3.0%

(b) Compound amount of positive cash flow when invested at 12%:
=[$20,000(P/A,12%,20) -$1,000(P/G,12%,20)](F/P,12%,20)
= $1,007,189
(F/P,i%,20) = $1,007,189/$170,750 = 5.899
ERR = 9.3%

(c) See part (c) of Problem C-1. 9.3% is a combination of 3% and 12% separate investments. The potentially separable decision as to the disposition of recovered funds, in this case, makes the project appear better than it is.

C-3

(a) Mere inspection of the cash flow streams shows clearly that Project B is preferable to Project A.

For A, NPW = 0 = -$100 +$65(P/A,i%,2) from which i = 20%
For B, NPW = 0 = -$100 +$55(P/A,i%,2) +$10(P/A,i%,10) from which i = 34%

(b) For Project A, F on date 2 with compounding at 4% = $133,673; solving for F/P of 1.3367 for 2 years gives an ERR of 15.6%. For Project B, F on date 10 is $275,011; solving for an F/P of 2.750 for 10 years gives an ERR of 10.6%.

(c) The use of the estimated terminal dates of the respective series of cash flows for different projects results in different percentages of dilution of the project so-called rate of return by the assumed reinvestment rate. Thus the 10 year Project B that yields 34% appears to yield only 10.6% because of the dominance of the 4% reinvestment return rate in the analysis method used by the WXY Company.

C-4

(a) NPW = 0 = -$100 +$19(P/A,i%,5) +$1(P/A,i%,20)
from which i = 3.5%

(b) FW of positive cash flows is:
$19(P/A,15%,5)(F/P,15%,20) +$1(F/A,15%,20)
= $1,144.844. Thus F/P = 11.448 from which ERR = 13.0%
In this case, the assumed reinvestment rate in the combination of investments increases the so-called rate of return from an unacceptable level to an acceptable one. One should question the use of a 15% reinvestment rate in combination with a much lower i* of only 10%.

218

PEE Solutions Manual   Appendix C

C-5

(a) $(F/P, i\%, 5) = 1.70$ from which ERR = <u>11.2%</u>

(b) Because all positive cash flow is at terminal date (date 5), no reinvestment at 15% enters into the calculation and i still appears to be <u>11.2%</u>

(c) No. The reason why Project E appears so good with this analysis method is that the 15% assumed reinvestment rate carries so much weight in the conclusion as to the project's merit. Even though the project itself will only yield 3.5%, the concentration of most of the positive cash flow in the early years in combination with the distant terminal date causes the computed so-called rate of return to be much closer to the 15% than to the 3.5%.

C-6

True rate of return on Q obviously is 50%.
For R, $(F/P, i\%, 20) = 3.50$ and rate of return is 6.46%.
Also, for (R-Q), $(F/P, i\%, 19) = 2.3333$ From which $i = 4.56\%$.
But if the XYW Company's system is used with a reinvestment at 4%, for Q, $F[20] = \$150,000(2.107) = \$316,000$; $(F/P, i\%, 20) = 3.16$ and the so-called rate of return appears to be about 5.9%, less than that of R.

The reinvestment assumption will have no influence on the overall valuation of R because there is nothing to reinvest. However, the reinvestment assumption reduces the overall return from Q from 50% to 5.9% because it creates a mixture of a 50% investment for one year with a 4% investment for the subsequent 19 years.

Project S obviously has a negative rate of return; half the invested funds are lost in one year, Project T has a value of $(F/P, i\%, 20)$ of 6.80 for which i is a little over 10%. But if the XYW Company's system is used with a reinvestment at 15% for S, $F[20] = \$50,000(14.232) = \$711,600$; $(F/P, i\%, 20) = 7.116$ and the so-called rate of return appears to be 10.3%, more than that of T.

The reinvestment assumption will have no influence on T because there is nothing to reinvest during the period of analysis. However, the assumption of a reinvestment at 15% for 19 years of the proceeds from the very unsuccessful investment in S actually makes S appear better than T.

The foregoing examples bring out the point that the difficulties in making sound comparisons using a reinvestment assumption, and thus a combination of separable investments, are not corrected by the choice of a common terminal date for all proposals.

PEE Solutions Manual   Appendix C

C-7

(a) The rate of return is found from:
NPW = 0 = -$10,000 +$2,800(P/A,i%,5) +$2,000(P/F,i%,5)
i = 16.5%

(b) Net annual return = $5,000 -$2,200
-($10,000 -$2,000)(A/F,10%,5) = $1,489.60
ERROR = $1,489.60/$10,000 = 14.9%

(c) Net annual return = $5,000 -$2,200 -$8,000(A/F,6%,5) = $1,380.80
ERROR = $1,380.80/$10,000 = 13.8%

(d) Net annual return = $5,000 -$2,000 -$8,000(A/F,16.5%,5)
= $1,648.17
ERROR = $1,648.17/$10,000 = 16.5%

While an interesting mathematical exercise, the argument is irrelevant. Depreciation, by any method, is not a cash flow. If invested capital is recovered, it is simply because the sum of the excess of receipts over disbursements, after consideration of income taxes, equals or exceeds the amount of capital invested. The disposition of any positive net cash flow is a separable decision and should be determined separately.

C-8

(a) The rate of return is found from:
NPW = 0 = -$10,000 +$1,600(P/A,i%,5) +$2,000(P/F,i%,5)
i = 0%; that is, [5($1,600) +$2,000 = $10,000]

(b) Net annual return = $3,800 -$2,200 -$8,000(A/F,10%,5) = $289.60
ERROR = $289.60/$10,000 = 2.9%

(c) The cash throw-offs are compound forward to time 5 as follows:
$2,000 +$1,600(F/A,10%,5) = $2,000 +$1,600(6.105) = $11,768
ERR = (P/F,i%,5) = $10,000/$11,768 = 0.8498
from which i = 3.3%

(d) Since the project returns nothing more than the original investment, the net return is explainable only as the result of the reinvestment of funds elsewhere at 10%. In (b), only the amount needed to recover the original capital investment is assumed to be reinvested. Neither method measures only the worth of the original investment proposal. Both involve the evaluation of a combination of investments.

PEE Solutions Manual   Appendix C

C-9

(a) Annual net return = $4,200 -$2,200 -$10,000(A/F,6%,10)
    = $1,241.30
    ERROR = $1,241.30/$10,000 = <u>12.4%</u>
    This would not be an attractive investment based on a
    company standard of 15%.

(b) The rate of return on this proposal is found from:
    NPW = 0 = -$10,000 +$2,000(P/A,i%,10)
    i = <u>15.1%</u>, therefore the proposal does in fact
    meet the company's standard of attractiveness, 15%.

(c) Annual net return = $4,200 -$2,200 -$10,000(A/F,15%,10)
    = $1,507.5
    ERROR = $1,507.5/$10,000 = <u>15.1%</u>
    This result is virtually the same as that in (b) because
    i* is so close to the true rate of return on the proposal.

(d) NPW = $2,000(P/A,15%,10) -$10,000
    = $2,000(5.019) -$10,000 = <u>+$38</u>
    Project is acceptable using this criterion.

(e) The solutions in (b) and (d) represent <u>only</u> the proposed
investment. By the choice of a MARR of 15%, the company presupposes that
any reinvested cash throw-offs (from whatever source) will earn 15% <u>or
more</u>. However, the method in (a) restricts a portion of the throw-offs
to earning only 6% throughout the life of the project. The <u>method</u>
assumes a compound investment of actually separable parts and thus the
method itself contains a conceptual error. This error in <u>concept</u> is not
overcome by using the MARR as the reinvesment rate as in part (c) even
though a correct decision could be made. The disposition of future cash
receipts is a future problem; a separable long range capital budgeting
problem.

C-10

(a) F = $2,000(F/A,6%,10) = $2,000(13.181) = $26,362
    (F/P,i%,10) = $26,362/$10,000 = 2.6362 from which
    ERR = <u>10.2%</u> Reject the project.

(b) From Problem C-9 (b), i = <u>15.1%</u>; and the project is acceptable.

(c) F = $2,000(F/A,15%,10) = $2,000(20.304) = $40,608
    (F/P,i%,10) = $40,608/$10,000 = 4.0608 from which
    ERR = <u>15.0%</u>

(d) From Problem C-9 (d), NPW = +$38; the project is acceptable.

(e) See part (e) of Problem C-9. The same conceptual error is being
made. The method understates the true rate of return even more than the
Hoskold-type method because all, rather than just a portion, of the cash
flow throw-offs are assumed to be reinvested.

PEE Solutions Manual

APPENDIX F

Depreciation Under the Tax Reform Act of 1986

General Notes:

Most of the problems in this Appendix have been solved using a personal computer spreadsheet because the depreciation charges under the Tax Reform Act of 1986 are neither uniform nor a gradient. The authors suggest that, whereever possible, the reader use a similar spreadsheet or a special financial calculator to save time in computing answers. Solving these problems using interest tables and an ordinary four-function calculator is very time consuming.

F-1 From Problem 9-29:
  (a) IRR B/T = 25.0%
  (b) IRR A/T = 16.3% (straight-line dep.)
  (c) IRR A/T = 17.0% (declining balance dep.)

Year	Cash flow before income taxes A	Write-off of initial outlay for tax purposes B	Influence on taxable income C	Influence of income taxes on cash flow -0.40C D	Cash flow after income taxes (A + D) E
0	($63,000)				($63,000)
1	16,300	($18,018)	($1,718)	$687	16,987
2	16,300	(12,852)	3,448	(1,379)	14,921
3	16,300	(9,198)	7,102	(2,841)	13,459
4	16,300	(6,552)	9,748	(3,899)	12,401
5	16,300	(5,481)	10,819	(4,328)	11,972
6	16,300	(5,481)	10,819	(4,328)	11,972
7	16,300	(5,418)	10,882	(4,353)	11,947
8	16,300		16,300	(6,520)	9,780
9	16,300		16,300	(6,520)	9,780
10	16,300		16,300	(6,520)	9,780
11	16,300		16,300	(6,520)	9,780
12	16,300		16,300	(6,520)	9,780
13	16,300		16,300	(6,520)	9,780
14	16,300		16,300	(6,520)	9,780
15	16,300		16,300	(6,520)	9,780
Sums	$181,500	($63,000)	$181,500	($72,600)	$108,900
		25.0% = IRR B/T		IRR A/T =	19.1%

PEE Solutions Manual   Appendix F

F-2

	Cash flow before income	Write-off of initial outlay for Tax	Influence on taxable	Influence of income taxes on cash flow	Cash flow after income taxes
year	taxes	purposes	income	-0.40C	(A + D)
	A	B	C	D	E
0	($18,000)				($18,000)
1	$6,500	($7,200)	($700)	$280	$6,780
2	$6,500	($4,320)	$2,180	($872)	$5,628
3	$6,500	($2,592)	$3,908	($1,563)	$4,937
4	$6,500	($1,944)	$4,556	($1,822)	$4,678
5	$6,500	($1,944)	$4,556	($1,822)	$6,688
5	$3,000	Sal. Value		($990)	
Sums	$14,500	($18,000)	$14,500	($6,790)	$10,710
	26.2%				18.3%

PEE Solutions Manual   Appendix F

F-3 (a) Applicable Tax Rate = (1 -0.05)0.33 +0.05 = 0.364 or 36.4%
       (b) (Headings same as in Problem F-2)

				.364D	
0	($60000)				($60000)
1	$15,000	($17,160)	($2,160)	$786	$15786
2	$15,000	($12,240)	$2,760	($1,005)	$13995
3	$15,000	($8,760)	$6,240	($2,271)	$12729
4	$15,000	($6,240)	$8,760	($3,189)	$11811
5	$15,000	($5,220)	$9,780	($3,560)	$11440
6	$10,000	($5,220)	$4,780	($1,740)	$8260
7	$10,000	($5,160)	$4,840	($1,762)	$8238
8	$10,000		$10,000	($3,640)	$6360
9	$10,000		$10,000	($3,640)	$6360
10	$10,000		$10,000	($3,640)	$6360
11	$10,000		$10,000	($3,640)	$6360
12	$10,000		$10,000	($3,640)	$6360
SUMS	$85,000	($60,000)	$85,000	($30,940)	$54,060

              19.7% = IRR B/T   IRR A/T = 14.9%

(c) Applicable tax rate = (1 -0.10)(0.33) +0.10 = 0.397 = 39.7%
    (Headings same as in Problem F-2)

				.397D	
0	($60000)				($60000)
1	$15,000	($17,160)	($2,160)	$858	$15858
2	$15,000	($12,240)	$2,760	($1,096)	$13904
3	$15,000	($8,760)	$6,240	($2,477)	$12523
4	$15,000	($6,240)	$8,760	($3,478)	$11522
5	$15,000	($5,220)	$9,780	($3,883)	$11117
6	$10,000	($5,220)	$4,780	($1,898)	$8102
7	$10,000	($5,160)	$4,840	($1,921)	$8079
8	$10,000		$10,000	($3,970)	$6030
9	$10,000		$10,000	($3,970)	$6030
10	$10,000		$10,000	($3,970)	$6030
11	$10,000		$10,000	($3,970)	$6030
12	$10,000		$10,000	($3,970)	$6030
SUMS	$85,000	($60,000)	$85,000	($33,745)	$51,255

              19.7% = IRR B/T   IRR A/T = 14.4%

PEE Solutions Manual   Appendix F

F-4 (a) (Headings same as Problem F-2)

	($60,000)			.68D $6,000	($54,000)
	$15,000	($9,231)	$5,769	($3,923)	$11,077
	$15,000	($8,462)	$6,538	($4,446)	$10,554
year	$15,000	($7,693)	$7,307	($4,969)	$10,031
0	$15,000	($6,924)	$8,076	($5,492)	$9,508
1	$15,000	($6,155)	$8,845	($6,015)	$8,985
2	$10,000	($5,386)	$4,614	($3,138)	$6,862
3	$10,000	($4,617)	$5,383	($3,660)	$6,340
4	$10,000	($3,848)	$6,152	($4,183)	$5,817
5	$10,000	($3,079)	$6,921	($4,706)	$5,294
6	$10,000	($2,310)	$7,690	($5,229)	$4,771
7	$10,000	($1,541)	$8,459	($5,752)	$4,248
8	$10,000	($754)	$9,246	($6,287)	$3,713
9	$85,000	($60,000)	$85,000	($51,800)	$33,200
10	19.7%	= IRR B/T		IRR A/T =	10.5%

(b) The increase in taxable income for Green does not change his tax bracket under the 1986 act. Therefore his after-tax rate of return is the same as in the Example F-1, 15.4%.

F-5 (Headings same as in Problem F-2)

0	($80,000)				($80,000)
1	$25,000	($32,000)	($7,000)	$2,800	$27,800
2	$30,000	($19,200)	$10,800	($4,320)	$25,680
3	$35,000	($11,520)	$23,480	($9,392)	$25,608
4	$40,000	($8,640)	$31,360	($12,544)	$27,456
5	$30,000	($8640)	$21,360	($8,544)	$21,456
Sums	$80,000	($80,000)	$80,000	($32,000)	$48,000
	26.9%	= IRR B/T		IRR A/T =	18.6%

PEE Solutions Manual   Appendix F

F-6 (Headings same as in Problem F-2)

Year						
0	($80,000)					($80,000)
1	$25,000	($53,360)	($28,360)	$11,344		$36,344
2	$30,000	($17,760)	$12,240	($4,896)		$25,104
3	$35,000	($8,880)	$26,120	($10,448)		$24,552
4	$40,000		$40,000	($16,000)		$24,000
5						
Sums	$50,000	($80,000)	$50,000	($20,000)		$30,000
	20.7%	= IRR B/T		IRR A/T =	15.3%	

F-7

Year	Cash flow before income taxes A	Write-off of initial outlay for tax purposes B	Influence on taxable income C	Influence of income taxes on cash flow −0.33C D	Cash flow after income taxes (A + D) E
0	($80,000)				($80,000)
1	25,000	($32,000)	($7,000)	$2,310	27,310
2	30,000	(19,200)	10,800	($3,564)	26,436
3	35,000	(11,520)	23,480	($7,748)	27,252
4	40,000	(8,640)	31,360	($10,349)	29,651
5		(3,000)	(3,000)	$990	990
6		(3,000)	(3,000)	$990	990
7		(2,640)	(2,640)	$871	871
Sums	$50,000	($80,000)	$50,000	($16,500)	$33,500
	20.7% = IRR B/T			IRR A/T =	14.9%

PEE Solutions Manual   Appendix F

F-8

Year	Cash flow before income taxes A	Write-off of initial outlay for tax purposes B	Influence on taxable income C	Influence of income taxes on cash flow -0.33C D	Cash flow after income taxes (A + D) E
0	($80,000)				($80,000)
1	25,000	($32,000)	($7,000)	$2,310	27,310
2	30,000	(19,200)	10,800	($3,564)	26,436
3	35,000	(11,520)	23,480	($7,748)	27,252
4	40,000	(8,640)	31,360	($10,349)	29,651
5	10,000	(8,640)	1,360	($449)	9,551
Sums	$60,000	($80,000)	$60,000	($19,800)	$40,200
	23.0% = IRR B/T			IRR A/T =	16.9%

PEE Solutions Manual   Appendix F

F-9 (a & b) Applicable tax rate = (0.908)(0.34) +0.092 = 40.0%

0	($240,000)				($240,000)
1	$50,000	($68,640)	($18,640)	$7,456	$57,456
2	$50,000	($48,960)	$1,040	($416)	$49,584
3	$50,000	($35,040)	$14,960	($5,984)	$44,016
4	$50,000	($24,960)	$25,040	($10,016)	$39,984
5	$50,000	($20,800)	$29,200	($11,680)	$38,320
6	$50,000	($20,800)	$29,200	($11,680)	$38,320
7	$50,000	($20,800)	$29,200	($11,680)	$38,320
8	$50,000		$50,000	($20,000)	$30,000
to					
15	$50,000		$50,000	($20,000)	$30,000
SUMS	$210,000	($240,000)	$210,000	($84,000)	$126,000
	19.4% = IRR B/T			IRR A/T = 15.0%	

(c)

Year	Cash flow before income taxes A	Write-off of initial outlay for tax purposes B	Influence on taxable income C	Influence of income taxes on cash flow -0.519C D	Cash flow after income taxes (A + D) E
0	($240,000)			$24,000	($216,000)
1	50,000	($30,000)	$20,000	($10,380)	39,620
2	50,000	(28,000)	22,000	($11,418)	38,582
3	50,000	(26,000)	24,000	($12,456)	37,544
4	50,000	(24,000)	26,000	($13,494)	36,506
5	50,000	(22,000)	28,000	($14,532)	35,468
6	50,000	(20,000)	30,000	($15,570)	34,430
7	50,000	(18,000)	32,000	($16,608)	33,392
8	50,000	(16,000)	34,000	($17,646)	32,354
9	50,000	(14,000)	36,000	($18,684)	31,316
10	50,000	(12,000)	38,000	($19,722)	30,278
11	50,000	(10,000)	40,000	($20,760)	29,240
12	50,000	(8,000)	42,000	($21,798)	28,202
13	50,000	(6,000)	44,000	($22,836)	27,164
14	50,000	(4,000)	46,000	($23,874)	26,126
15	50,000	(2,000)	48,000	($24,912)	25,088
Sums	$510,000	($240,000)	$510,000	($240,690)	$269,310
	19.4% = IRR B/T			IRR A/T =	13.7%

PEE Solutions Manual   Appendix F

F-10 5-Year Class Depreciation

0	($ 8,400)				($ 8,400)
1		($ 3,360)	($ 3,360)	$ 1,176	$ 1,176
2		($ 2,016)	($ 2,016)	$ 706	$ 706
3		($ 1,210)	($ 1,210)	$ 484	$ 484
4		($ 907)	($ 907)	$ 363	$ 363
5		($ 907)	($ 907)	$ 363	$ 363
6			$ 0	$ 0	$ 0

                                          NPV =   ($ 5,804)
                                          EUAC A/T = $ 1,333

Straight-Line Depreciation

0	($ 8,400)				($ 8,400)
1		($ 3,360)	($ 1,400)	$ 490	$ 490
2		($ 2,016)	($ 1,400)	$ 490	$ 490
3		($ 1,210)	($ 1,400)	$ 560	$ 560
4		($ 907)	($ 1,400)	$ 560	$ 560
5		($ 907)	($ 1,400)	$ 560	$ 560
6			($ 1,400)	$ 560	$ 560

                                          NPV =   ($ 6,481)
                                          EUAC A/T = $ 1,488

PEE Solutions Manual   Appendix F

F-11

		Before 1986 Act		After 1986 Act
	Book Val			
Year	end of yr.	Depreci.	Percentage	Percentage
0	$1,000.00				$161.89
1	$715.00	$285.00	28.50%	28.57%
2	$511.23	$203.78	20.38%	20.40%
3	$357.23	$154.00	15.40%	14.58%
4	$203.23	$154.00	15.40%	10.41%
5	$49.23	$154.00	15.40%	8.67%
6				8.67%
7				8.67%
SUM		$950.78	95.08%	99.97%

F-12

Most small businesses either lose money the first year or two or make very little. The use of the 100% write-off of $10,000 investment in qualifying asset would be of little or no value in those years. First, the tax rate would be zero or very low. Second, as the business prospers, the tax rate bracket will be higher. Thus write-off by the 7-year class depreciation schedule or the straight-line method might produce greater overall tax benefits.

230